NORTHLAKE PUBLIC LIBRARY DIST

P9-DMJ-781

3 1138 00168 4852

THE TOXIC SANDBOX

DATE DUE

THE TOXIC SANDBOX

THE TRUTH ABOUT ENVIRONMENTAL
TOXINS AND OUR CHILDREN'S HEALTH

Libby McDonald

A Perigee Book

A PERIGEE BOOK
Published by the Penguin Group
Penguin Group (USA) Inc.
375 Hudson Street, New York, New York 10014, USA

Penguin Group (Canada), 90 Eglinton Avenue East, Suite 700, Toronto, Ontario M4P 2Y3, Canada (a division of Pearson Penguin Canada Inc.) • Penguin Books Ltd., 80 Strand, London WC2R 0RL, England • Penguin Group Ireland, 25 St. Stephen's Green, Dublin 2, Ireland (a division of Penguin Books Ltd.) • Penguin Group (Australia), 250 Camberwell Road, Camberwell, Victoria 3124, Australia (a division of Pearson Australia Group Pty. Ltd.) • Penguin Books India Pvt. Ltd., 11 Community Centre, Panchsheel Park, New Delhi—110 017, India • Penguin Group (NZ), 67 Apollo Drive, Rosedale, North Shore 0632, New Zealand (a division of Pearson New Zealand Ltd.) • Penguin Books (South Africa) (Pty.) Ltd., 24 Sturdee Avenue, Rosebank, Johannesburg 2196, South Africa

Penguin Books Ltd., Registered Offices: 80 Strand, London WC2R 0RL, England

While the author has made every effort to provide accurate telephone numbers and Internet addresses at the time of publication, neither the publisher nor the author assumes any responsibility for errors, or for changes that occur after publication. Further, the publisher does not have any control over and does not assume any responsibility for author or third-party websites or their content.

Copyright © 2007 by Libby McDonald
Cover art by Gerry Images
Cover design by Ben Gibson
Text design by Tiffany Estreicher

All rights reserved.
No part of this book may be reproduced, scanned, or distributed in any printed or electronic form without permission. Please do not participate in or encourage piracy of copyrighted materials in violation of the author's rights. Purchase only authorized editions.
PERIGEE is a registered trademark of Penguin Group (USA) Inc.
The "P" design is a trademark belonging to Penguin Group (USA) Inc.

First edition: November 2007

Library of Congress Cataloging-in-Publication Data

McDonald, Libby.
    The toxic sandbox : the truth about environmental toxins and our children's health / Libby McDonald.—1st ed.
        p. cm.
    "A Perigee book."
    Includes bibliographical references.
    ISBN-13: 978-0-399-53363-1 (Perigee trade pbk.)
    1. Pediatric toxicology—Popular works.   2. Environmental toxicology—Popular works.
I. Title.
    RA1225.M43 2007
    618.92'98—dc22

                                                                    2007026768

PRINTED IN THE UNITED STATES OF AMERICA

10 9 8 7 6 5 4 3 2 1

Most Perigee books are available at special quantity discounts for bulk purchases for sales promotions, premiums, fund-raising, or educational use. Special books, or book excerpts, can also be created to fit specific needs. For details, write: Special Markets, Penguin Group (USA) Inc., 375 Hudson Street, New York, New York 10014.

For David, George, Daniel, Matilda, and Paulina

8-14-08

NORTHLAKE PUBLIC LIBRARY DIST
231 N. WOLF ROAD
NORTHLAKE, IL 60164

CONTENTS

FOREWORD

In recent years research scientists, health care professionals, and the general public have become increasingly concerned that exposure to chemicals in the environment or
various consumer products may be harmful. These concerns are well founded. But they have also caused confusion and even a sense of helplessness. Parents often wonder,
"What am I supposed to do? I'm not a scientist. Isn't some-
one in the government paying attention?"

A number of years ago, as a practicing physician, I also
began to wonder about the health impacts of environmental contaminants. My medical school training was of little
help. It included virtually nothing about these concerns.
But that was when Cleveland's Cuyahoga River caught fire
near my school and only a few years after Rachel Carson

published *Silent Spring*. The public environmental health impacts of a rapidly growing chemical industry were not yet on the radar screen of the general public and only beginning to attract the attention of thoughtful scientists. Most of us were basking in the glow of promises of "Better Things for Better Living . . . Through Chemistry."

Decades later I studied the history of the pesticide DDT in the United States. It is an instructive and cautionary tale. Widespread use of DDT in agriculture after World War II led to contamination of humans, wildlife, and ecosystems around the world. But the story of attempts to restrict its use is one of sharp disagreements over scientific data, scientific uncertainty, and contested efforts to protect public health and the environment. Science, values, and economic and political power intertwined in acrimonious courtroom hearings. Ultimately, the newly formed Environmental Protection Agency took on pesticide regulation and in the 1970s banned DDT because of its role in decimating bird populations and its suspected impacts on people. But the furor over DDT regulation was only a preview of today's highly politicized governmental oversight of the manufacture, use, and disposal of industrial chemicals.

Many studies show that some communities and individuals are exposed to unsafe levels of dangerous pesticides that drift in the air, onto playgrounds, and into homes, where they contaminate children's toys, furniture, and house dust. Moreover, results of recent scientific studies lend support to concerns about the impacts of low levels of some pesticide exposures on a developing child's brain and immune system, as well as on cancer risk.

But pesticides are only part of the problem. The vast majority of thousands of other industrial chemicals are far less regulated. Chemicals that are not pesticides or pharmaceuticals are not required to undergo any premarket testing before they end up in consumer products or otherwise find their way into air, water, soil, or food. In fact, when such a chemical is proposed for marketing, the administrator of the Environmental Protection Agency must be able to justify concerns about its toxicity or conclude that significant exposures are likely before the agency is authorized to require premarket testing. Of course, without safety data, it's difficult to justify the concern—a catch-22 that rewards ignorance and lack of information. To make matters worse, thousands of chemicals that were already on the market when current regulatory laws were passed remain on the market without comprehensive safety data. The burden of proof remains on the government to demonstrate a problem—not on the manufacturer to assess safety.

Recently, the European Union adopted a more sensible chemical regulatory policy. Known as Registration, Evaluation, and Authorization of Chemicals (REACH), this law requires more comprehensive safety evaluation of chemicals proposed for marketing, safety testing of chemicals already on the market or their withdrawal after a certain period, and safer substitutes for the most hazardous chemicals. As it made its way through a complex political process, the proposed law was subject to intense efforts to defeat it by those whose primary interest was to protect the economic interests of chemical manufacturers—including the US Department of Commerce and Department of State in the Bush administration.

Many people are hopeful that more protective chemical policies will ultimately be adopted in the United States. Several states have adopted legislation that addresses specific chemicals of concern and there are early glimmers of more comprehensive efforts. But we have a long way to go before state and federal governments are truly meeting their public trust responsibilities to keep hazardous chemicals out of developing babies, breast milk, and the blood of most Americans.

This is the world that Libby McDonald encountered when she plunged into the task that resulted in this book. She came through a well-traveled doorway—motivated by a mother's love and concerns about the health of her children. But Libby didn't stop with her own questions. She learned that other parents had similar concerns but few answers. Drawing on her curiosity, considerable skills, and devotion to children, Libby has produced a remarkably useful and eminently understandable book. It will be invaluable for parents and others who have had similar questions and want help sorting out information before deciding what to do.

The scope of what Libby McDonald has accomplished

should not be underestimated. Beginning as a concerned mother, not as a scientist, she set out to become responsibly informed before deciding what kind of advice to offer others. Her journey is documented in the pages that follow. In the process, Libby shows the important role that each of us plays in influencing health-protective public policy—whether in homes, schools, communities, or governments. She connects the dots between individual purchasing decisions and their far-reaching influences on the health of many children and the quality of environments many miles away. She shows how political and economic decisions concerning electricity or food production or city planning can influence the ability of a child to learn and remember.

This book offers straightforward advice based on the synthesis of information from many sources. It is a generous gift to all people who want what is best for children but sometimes don't know where to begin when faced with confusing and often conflicting science. It may even spark curiosity in others who will undertake their own journeys because of heartfelt concerns for the health and well-being of this and future generations. It is an important contribution for which many will be grateful.

—Ted Schettler, M.D., Science Director,
Science and Environmental Health Network

1

UNCOVERING
THE TRUTH

A YEARLONG INVESTIGATION

To be honest, before I had my own children I didn't think a whole lot about the environment. I blocked out the constant back-and-forth between environmental types and their critics. The shrill tone of it all, the shifting issues—I could not be bothered. Then, after I had my first baby, the debate crept into my mind. Glancing at old peeling paint or a slab of barbecued salmon, my head would race, remembering an article I had read weeks ago, tipping me off to a particularly vile toxin that wreaks havoc on a child's nervous system.

As I had my second and then my third child, more and more newspaper headlines began to report connections between environmental threats and mushrooming childhood ailments. I was less successful at blocking out the noise and experienced moments of dread that I wasn't sufficiently protecting my children from serious environmental contaminants. Even so, because it is downright scary to think about toxins' reported side effects—learning disabilities, asthma, cancer—I continued to try my best to put it out of my mind, telling myself that not only were the media reports overblown, but there was nothing I could do about the toxins that supposedly contaminated my children's food, air, and water.

Then, my two-year-old had a high lead count and unfortunately I wasn't the only one whose child seemed suddenly at risk. Just in my circle of close friends, as our children became toddlers and went off to school, one parent's little girl was diagnosed with a mild form of autism where she had difficulty walking and suffered seizures. Another learned her seven-year-old had attention deficit hyperactivity disorder (ADHD). He was asked to leave the school he had attended since he was four. I began to wonder if there was a connection between these chronic and serious ailments and environmental toxins.

Childhood illness has changed dramatically in the last one hundred years. No longer do children in industrialized countries die of infectious diseases like polio, measles, or meningitis. That is the good news. The bad news is more and more of our children suffer a host of chronic conditions scientists call the "new pediatric morbidity." These ailments include asthma, cancer, ADHD, poor motor skills, preterm birth, autism, and some birth defects.

At the same time that these chronic conditions have multiplied in the last few decades, becoming alarmingly more common, some eighty thousand new synthetic chemical compounds have been introduced into the environment. Public health scientists have responded by aggressively examining the link between toxins and children's health; and in 1996 the US Environmental Protection Agency (EPA) established the Office of Children's Health Protection.

Are these chemicals finding their way into our children's bodies? In 2004 the Red Cross decided to take a look at umbilical cord blood and figure out what chemicals were pumped across the placenta to our children in the womb, the most vulnerable time of their development. By analyzing the cord blood of ten newborns, the researchers learned that the babies in utero had an average of two hundred industrial chemicals and pollutants running through their veins. In total 287 toxins were identified in the ten babies tested, pollutants that are found in pesticides, stain repellents, flame retardants, waste from coal-burning power plants, gasoline, and garbage. Of these chemicals 180 are known to cause cancer, 217 are poisonous to the brain and nervous system, and 208 have been linked to birth defects in animal studies.

While it is clear that harmful pollutants and chemicals are in our children's bodies, the question is what amount of exposure is dangerous. Researchers are constantly uncovering new information, making it hard for medical professionals and government agencies to catch up with cutting-edge findings. Furthermore, given the track record of business and the US government (the lead industry refused to acknowledge that lead was hazardous to children years after public health scientists, in multiple studies, had demonstrated that it was), it is nearly impossible for a parent to feel confident that these entities are sufficiently looking out for the welfare of our children.

Long after the surgeon general's 1964 report linked cigarette smoking to lung cancer, the tobacco industry maintained that cigarettes were not nearly as dangerous as scientists claimed, nor were they addictive. Today, not only do we recognize that cigarettes are responsible for the deaths of millions of smokers and that you can easily get hooked, but we now know that nicotine crosses the placenta and, during key moments of brain development, can harm the developing fetus, resulting in learning and behavioral problems as well as increased asthma symptoms.

Then, just when we thought we knew everything about the perils of smoking, a 2005 study yielded a significant finding: that nicotine crosses generations, injuring future offspring. The bewildering results showed that if a child's grandmother smoked while pregnant with the child's mother, the child has twice the risk of developing asthma. Researchers suspect a mother's smoking actually turns on or off genes in the mitochondrial DNA, resulting in future generations' diminished immune function and greater susceptibility to asthma. The ability of a toxin to actually change the way genes express themselves is significant, for it means the toxin not only harms our children but it alters the genes they pass down to our grandchildren.

Our understanding of the effects of alcohol on the developing fetus has also changed dramatically in the last fifty years. When my mother was pregnant with my oldest sister in the late 1950s she attended a natural childbirth class. Careful not to exceed guidelines set by her obstetrician, each night before heading off to learn about giving birth without the use of drugs, she and my father would have a martini. Nearly forty years later when I was pregnant with my first child, Matilda, my obstetrician advised that I drink no more than one glass of wine a week. However, in 2003, when I was carrying my youngest, George, I was told to avoid alcohol altogether. During the span of thirteen years between my oldest and youngest child, scientists had learned that alcohol is far more toxic to babies in the womb than we had once thought.

In the end, this book was not inspired by my grave concern about rising rates of childhood ailments. Nor did I write the book as a reaction to a steady supply of anxiety-provoking media headlines reporting everything from smaller penises to plummeting IQs. In the end, my submersion into the world of children's environmental health was the result of my own experience with my youngest child's elevated lead count.

What was most significant about George's brush with lead is that even with a neurotoxin whose effects on children have been studied for nearly one hundred years, pediatricians, government agencies, and brain specialists all gave me—the mother—very different information. Initially, as required by New York State law, George was tested for lead, and when I asked about his lead level, the pediatrician assured me that it was okay. Six months later, at George's next checkup, I glanced at his chart and was alarmed when I read that his lead count was 10 μg/dl (the number refers to micrograms per deciliter) on the scale rating toxicity. I questioned the doctor about the implications of a lead count of ten and with real certainty in her voice she told me that the US Centers for Disease Control and Prevention (CDC) considered his lead level safe.

Even though my children's pediatrician had emphasized that there was nothing to worry about, anxiety over that amount of lead pulsing through my baby's veins kept gnawing at me. That night, after getting the children to bed, I searched the Internet, surfing environmental websites for information about lead poisoning. Although the pediatrician was right about threshold levels established by the CDC, what I discovered was so scary I began to fantasize about packing up the children and running far away into the mountains, distancing my little ones from our tainted civilization.

Then, three days later at a dinner party, by chance I was seated next to a pediatrician who was also a lead expert. A thin, meticulously groomed woman in her mid-forties, she matter-of-factly stated that the latest research shows that anything above 2 µg/dl can be associated with a lower IQ, anemia, and behavioral problems. Frantic to find the source of George's lead contamination, I became suspect of everything in my home, giving away valuable antiques regardless of their price, obsessively sweeping up dirt tracked into the foyer, and chucking an entire set of Italian painted dishes.

Soon thereafter, my husband wondered out loud about the clue "It has lead in it" in the *New York Times* crossword puzzle. I scanned the letters already filled in— "p - - - er"—and felt sick. I had been spooning cereal laced with traces of lead into George's gaping mouth from an antique pewter bowl.

Why wasn't George's pediatrician aware of the latest research on this nasty neurotoxin? A 2006 study that mailed questionnaires to pediatricians showed that although doctors are aware that many major chronic illnesses today's children suffer from—including asthma, autism, attention deficit disorders, and even cancer—can be exacerbated by or even directly result from exposure to environmental contaminants, most doctors have no training in environmental health. Furthermore, except for lead, they haven't been taught to gather information about a child's environmental history and have little to no ability to talk about threats posed by environmental toxins like mercury, plastics, and pesticides. With 93.5 percent of the New York State participants reporting that they've had a patient harmed by exposure to an environmental toxin, it is no wonder that pediatricians believe environmentally related sicknesses in children are growing.

Today in the Adirondacks, the climate is changing drastically. Our oldest child, Paulina, who is seventeen, intended to teach skiing at Whiteface Mountain so that she could earn enough money to travel to Ireland after graduating from high school. The temperature, which in years gone by never climbed above freezing until late March, has been in the fifties—not nearly cold enough for making snow. Indeed, the problem of climate change overshadows concerns of environmental contaminants and children's health. However, solutions are closely knit together. For instance if you choose to buy locally grown organic food you are not only eliminating trace amounts of chemicals from your children's diet but you are helping reduce the fog of pollutants that are released into the air and contribute to global warming when trucks transport produce from hundreds of miles away to your local Piggly-Wiggly.

Although George is now smart, coordinated, and (with an enthusiastic audience of three older siblings) a ham who demands to be called "Super George," his ordeal with lead illustrates that no matter how vigilant we are, environmental toxins can still creep into our homes. Coupling that with a newfound certainty that environmental exposures can make our children sick, I decided to plunge into the world of children's environmental health. In the end, *Toxic Sandbox* is the story of my journey to unearth the key toxins that threaten the well-being our children. I interviewed experts in the fields of medicine, anthropology, zoology, education, and public health; visited families who suspected their children had been harmed as a result of environmental contaminants; mailed off samples of cosmetics, soil, hair, and paint chips to be tested for toxicity. As I pored over studies that offer evidence about how toxins have been shown to affect children, I learned prudent, easy strategies we can all use to protect our kids from environmental illness in the twenty-first century. Whether it is read from front to back, or used as a resource guide, such as for information on a toxin of particular concern, *Toxic Sandbox* is ultimately an easily accessible manual for parents, health professionals, teachers, students, school administrators, community activists, and lawmakers to finally get real answers not only about the link between children's health and environmental risks but about the steps we can take to keep our children safe from environmental threats.

2

LEAD

THERE IS *NO* SAFE THRESHOLD

Even though the dangers of lead have been known for decades, it is the first thing to worry about. Cutting-edge research shows that our children's brains are much more sensitive to this metal than we previously thought. Widely used for much of the twentieth century in gas and paint, lead is still pervasive. It is my recommendation that if you live in a house built before the 1978 ban on lead paint, get your children tested for lead and be extremely cautious when doing even the simplest renovations on your home.

THE BAD NEWS

- The CDC says that one-quarter of US children are exposed to lead in their own homes; and with 80 percent of American houses built before the 1978 ban on lead paint, the threat of exposure will not go away for a long time.

- There is mounting evidence that there is no safe threshold for lead in the body—evidence that conflicts sharply with the CDC's position that up to ten micrograms per deciliter is acceptable.

- Exposure to lead in early childhood causes lifelong impairments for which there are no simple treatments.

- Lead in children is associated with:
  - A loss of IQ points
  - Reading problems
  - Failure to grow
  - Hearing loss
  - Speech deficits
  - Attention deficits
  - Antisocial behavior
  - Aggressive behavior
  - Delinquency and criminal behavior

THE GOOD NEWS

- Blood lead levels have dropped dramatically since the government banned lead in paint (1978) and lead in gasoline (1991). According to the National Health and Nutrition Examination Surveys (NHANES), the average blood lead level in children has come down by approximately 80 percent since the 1970s. Today, the average count is somewhere between 1.5 and 2 micrograms per deciliter.

- As a child your lead levels were probably higher than those of your own children's lead levels and you turned out okay, didn't you? Even so, today we have a clearer understanding of the effects of lead, and there are a number of easy and straightforward steps you can take to ensure that your child's lead count stays as close to zero as possible.

THE SCIENCE

Study: Silent doses of lead injure children

To understand how lead affects a child's health, in 1976 Dr. Herbert Needleman of the Boston Children's Hospital asked 2,500 students from twenty-nine schools in neighboring towns to donate their baby teeth for a study. Since lead levels are easily read in teeth, Dr. Needleman's collection allowed him to compare each child's school progress with his or her lead exposure. What he learned is that higher lead levels directly correlated with lower IQ scores, decreased attention span, and poor language ability. In 1988 Dr. Needleman contacted half of the children who'd contributed to the original study and found that those who had higher lead levels were seven times less likely to graduate from high school and six times more likely to have a reading disability.

Study: Even the tiniest exposure to lead harms children

To make things even scarier, a 2005 study by Dr. Bruce Lanphear, the Sloan Professor of Children's Environmental Health at Cincinnati Children's Hospital Medical Center, showed that lead harms children at levels considerably lower than the CDC's threshold of ten micrograms per deciliter. It may seem strange but there is a steeper drop in intellectual ability when the child's level of exposure

Children show a decline of 7.4 IQ points for the first ten micrograms of lead per deciliter of blood. There is a loss of 4.6 IQ points for every ten micrograms thereafter.

goes from five to ten micrograms per deciliter than there is when it goes from ten to twenty.

"Isn't lead and children's health old news?"

Lead counts have fallen dramatically over the last twenty years. However, new research shows that even *tiny* amounts of lead can damage a child's developing brain, resulting in lasting effects on his or her ability to think and concentrate. The equation is simple: when lead counts go up, reading scores and overall classroom performance go down.

"DALLY NO LONGER: GET THE LEAD OUT"
—*New York Times* (January 17, 2006), F6

"Why are children particularly vulnerable to lead?"

The body has a built-in mechanism for protecting the brain from poisonous substances, called the blood-brain barrier (BBB). Unfortunately, lead befuddles the BBB by masquerading as calcium, which crosses the barrier easily. Because babies' and toddlers' brains crave calcium in their effort to develop, until children are around six years old they can absorb three to four times more lead than an adult. And if a child has a calcium deficiency it makes matters worse: his or her developing body will grab on to this calcium look-alike, accelerating the uptake of lead. During this process, of course, children's brains are rapidly developing and are particularly sensitive to the effects of lead.

Q&A:
Dr. Bruce Lanphear

Q: Should we make doubly sure kids get enough calcium as a safety measure?

Dr. Lanphear: There is convincing evidence from animal studies that calcium supplementation will reduce lead absorption, but in human studies, the evidence isn't as clear-cut. If calcium supplementation does reduce lead absorption in children, it is clear that it is not sufficiently protective (i.e., the reduction of blood lead level is small).

"Can my unborn baby be exposed to lead?"

Lead also passes through the umbilical cord to our unborn babies. Although most women in this country have low lead counts, even grownups can be seriously exposed to lead by, say, living through the renovation of an old house or drinking water that has traveled through lead pipes. If you live in a home built before 1978, when lead was finally banned, you may want to request a lead test before becoming pregnant.

However, there is another, more concerning way lead can enter a mother's bloodstream. Half of the lead ingested or inhaled when we were girls gets stored in our bones for up to thirty years. Then during pregnancy and lactation, when our bodies require additional calcium, the lead from our bones activates and enters our bloodstream. A calcium supplement during pregnancy may prevent the transfer of lead to our unborn babies.

Study: Early lead exposure stored in bones activates during pregnancy

The chemical fingerprint of lead in people living in Australia is actually different from that found in people of other countries. For this reason, Dr. Brian Gulson of Macquarie University in Sydney, Australia, was able to analyze the composition of lead in immigrants' blood and use the distinction between outside lead and Australian lead as a data point for determining lead exposure prior to arrival in the country. He learned that the lead that mobilized from their bones into their bloodstream during pregnancy was the lead women ingested early in their lives, before they came to Australia. The amount released during the postpartum period measured anywhere between 0.9 and 10.1 micrograms per deciliter. In other words, pregnancy doubled the lead levels in the blood of women who participated in the study. The results from this study are alarming because they show that lead stored in a mother's bones when she was a girl can actually pass to her baby in utero.

Q&A: Dr. Brian Gulson

Q: Do calcium supplements keep lead from mobilizing from our bones during pregnancy and lactation?

Dr. Gulson: The calcium supplements appear to assist in reducing the lead mobilization during pregnancy and a pregnant women would commonly be taking a calcium supplement anyway—but calcium supplementation does not help during lactation.

"I'm still nursing my eight-month-old and I just had my blood tested for lead—my lead count was five. What does this mean for my nursing baby?"

Lead can also find its way into the most sacred of nutrients: our breast milk.

Nonetheless, all the experts agree that breastfeeding should continue through age one unless a mother was severely lead poisoned at some time in her life. Dr. Brian Gulson explains, "We have shown that the transfer of lead to the infant from breast milk is low." If you are concerned, ask your doctor to test your blood for lead. If your blood lead level is less than ten micrograms per deciliter, Dr. Gulson assures that there is no need for concern, as the amount of lead in breast milk is probably at the most only about 5 percent of that in your own blood.

"How are children tested for lead exposure, and if they have an elevated lead count, how does a specialist determine whether or not they have been critically harmed?"

Lead screening requires drawing blood from the vein. Finger pricks, done in some schools and pediatricians' offices, do not yield enough blood for really accurate testing. Take note that many pediatricians are not aware of the new research showing how very low levels of lead can injure a child's brain. For this reason, ask for the exact count, measured in micrograms of lead per deciliter of blood.

Dr. Theodore Lidsky, both a neuropsychologist and a toxicologist with the New York State Institute for Basic Research in Developmental Disabilities in Staten Island, evaluates children who have been exposed to lead. "Lead causes brain damage," says Dr. Lidsky. "So I evaluate a child with an elevated lead count much the same way I would a child whose head has gone through a windshield." To explore how a child's brain has been injured, Dr. Lidsky uses IQ testing and nueropsychological testing. He explains that lead targets developing brain cells. Because the frontal lobe—responsible for planning, judgment, and concept formation—is the last part of the brain to fully develop, it tends to be at the greatest risk for lead poisoning.

It is important to note that all brain damage caused by lead has a lag effect. Often with lead, you will not see the deficits until a child is six or seven. Or even, at times, as late as his or her early teen years. What this means is that if a child has a high lead count at one or two, everything may appear to be fine until age six or seven, when the region of the damaged brain actually begins to develop. Dr. Lidsky advises parents who suspect their child has been exposed to lead to have the child screened. He goes on to say, "If I had a child with a blood lead level of four or five, I would give them calcium and find them a pediatrician who understands lead."

Study: Attention deficit hyperactivity disorder (ADHD) can be attributed to lead exposure

In 2006 Dr. Lanphear published a study in *Environmental Health Perspectives* that attributes about one-third of attention deficit disorder among US children to tobacco smoke before birth or to lead exposure in early childhood. The study analyzed data on 4,704 children ages four to fifteen

---

Today an estimated 310,000 US children ages one to five have lead levels exceeding the government's "acceptable" blood lead level of ten micrograms per deciliter.

---

with concentrations of lead in blood samples and prenatal exposure to cigarette smoke. Children with blood lead levels of more than two micrograms per deciliter were four times more likely to have ADHD than children with lower levels.

"Where are our children exposed to lead?"

Children are exposed to lead in two ways: they either swallow it or they breathe it in. It is nearly impossible to absorb lead through the skin. The most common ways children are exposed to lead include house paint, soil, plumbing, food containers, and various other products and items.

LEAD PAINT IN OLD HOMES. Lead in paint was discovered to be dangerous as early as 1900 and was banned by many countries as part of an international convention in 1925. However, much like the Kyoto Accord today, the United States did not join the worldwide effort to eliminate lead paint back then; it waited until 1978 to ban it. Today, lead paint in an old home is the most common way for a child to get exposed to lead. It is estimated that 42 to 47 million homes in this country still contain lead paint, mostly in the eastern United States, where parts of New York City are considered to be the lead belt of the country. In apartments and rental properties where tenants frequently come and go, it is more likely that paint is chipping and peeling, making these properties a greater threat to children. Furthermore, lead paint is sweet, sometimes enticing children who have gotten a taste of it to consume more.

"My husband is really into This Old House *and is constantly updating our hundred-year-old home. Could he possibly be exposing our young children to lead in paint?"*

Time and again my research has turned up stories of high lead levels in children whose parents are fixing up their old houses. Anytime anyone does renovations in your home, they may release lead dust, even if it has been safely enclosed by layers of more recent paint. Once lead is in the air, your children can be exposed by breathing it in.

SOIL NEAR ROADS, OLD HOUSES, AND FENCES. In the early 1920s a chemist at the General Motors Company named Thomas Midgely discovered an effective antiknock agent by adding tetraethyl lead to gas. Even though workers at three factories that produced the new additive became psychotic, and two workers died, leaded gas was found safe by the US government and first went on sale in the United States in 1923. Between 1976 and 1994, as leaded gas was phased out and eventually banned, the NHANES determined that the average blood lead level dropped from 13.7 micrograms per deciliter to 3.2 micrograms per deciliter. Even so, lead lives on in dirt alongside roadways as well as in soil around old houses and fences where lead paint, over the years, has chipped away. Today this is how some US children are exposed to lead: they play outside near old homes or close to roads, put their hands in their mouths, and swallow lead in the dirt.

PLUMBING IN OLD HOMES. Some of the nation's drinking water still travels through lead pipes and lead solder into our homes. This is particularly true in older communities, where houses may connect to the drinking water system through lead pipes that predate a twenty-year-old revision of the Safe Drinking Water Act. To test your water call the US EPA Drinking Water Hotline and locate a local certified testing laboratory. Read the label when buying a water filter. You must use a filter that is effective at removing metals. Lead is not a problem when showering because it is not absorbed through the skin.

FOOD CONTAINERS. Not only is lead solder sometimes used to seal imported canned food, but it can leach into food served on dishes painted with lead paint or from old pewter that is made from an amalgam of metals including lead.

FOLK REMEDIES. Some imported traditional medicines, including many from southern Asia, contain lead.

TOY JEWELRY, LUNCH BOXES, FLASHLIGHTS, FISHING RODS, LIPSTICKS, HAIR DYES, CALCIUM SUPPLEMENT TAB-LETS. Last year the Consumer Product Safety Commission recalled over 150 million toys that contained lead, and instituted more stringent policies regarding lead in children's jewelry. To stay on top of product recalls sign up for the CPSC e-mail list at www.cpsc.gov/cpsclist.aspx.

Real Stories

In February 2006 a four-year-old boy was brought to a hospital emergency room in Minneapolis with vomiting. His parents were told to increase fluids and they took him home. Two days later, still listless, he was again at the ER with more vomiting and an upset stomach. Ten hours after being admitted the boy became agitated and had a seizure. A computer tomography scan showed a heart-shaped object in his belly. When he was tested for heavy metals his doctors learned that he had a lead count of 180. Four days after being admitted the boy died. His autopsy revealed that the heart-shaped charm was stamped with the insignia of a popular sneaker brand. His mother recognized it from a new pair of shoes that belonged to his friend. The Minneapolis Public Health Department determined that the ingested charm consisted of 99.1 percent lead.

"My husband and I want to buy an old home. What should we know about possible lead hazards?"

Under federal law, the 1992 Residential Lead-Based Paint Hazard Reduction Act requires people selling their home to disclose any information they have about lead paint in their house. However, they are not required to actually gather any such information by, say, testing their home for lead before putting it on the market. This offers little protection to buyers—as long as the seller remains in the dark about lead in his or her house, there is nothing that needs to be disclosed. Therefore you must check for yourself.

When considering a new home, you should examine all interior and exterior surfaces for peeling, chipping paint, or alligatoring (when old paint resembles alligator skin). Pay especially careful attention to window frames and radiators. To get a general idea of whether or not lead is even an issue, you can use a lead-testing pen that is available at many hardware stores (however you must get down to the lowest layer of paint to make an accurate assessment). A certified environmental technician can perform more accurate lead testing using X-ray fluorescence—a sort of wand that is passed over surfaces. You should be able to find a certified lab by calling the National Lead Information Center at 1-800-424-LEAD.

Some counties offer financial assistance for removing lead from homes. Be extremely careful: when done wrong, lead abatement can make things worse by not properly containing lead dust generated in the process. The dust can invisibly settle on nearby surfaces and later be unknowingly inhaled or ingested.

"KEEP LEAD AND ARSENIC IN THE GROUND AND
OUT OF YOUR MOUTH"
—*Tacoma News Tribune* (July 16, 2005), E1

"I live in an old Victorian and I do a lot of gardening in beds next to our house. What should I know about lead-contaminated soil?"

Select a site as far away from city streets and old buildings as possible. If you suspect that there is lead in your soil, get it tested. Although lead has been known to collect in the roots of plants, it does not appear to move into shoots. For this reason avoid planting root crops like potatoes, beets, and carrots in lead-contaminated soil.

Real Stories

In 2004, when Olin was still crawling, his parents realized their dream when they bought an old camp in the Hudson Valley and began renovating one of many buildings on the twelve-acre parcel of land. Olin's mother, a tall, thoughtful woman in her midtwenties, recounts, "There was a lot of paint flaking off, but at the time we weren't even thinking about lead. We tore down the entire porch and I always had him on my back." At age one, as is required by law in New York State, Olin was tested for lead and the results came back at twenty-seven micrograms per deciliter. His mother explained, "It was so abstract and surreal because I realized he had been injured but I couldn't see it."

The health department came to visit their home to determine the source of Olin's lead exposure. They found high counts of dust outside the home and in the ambient air inside and concluded that dust from the demolition work had been tracked in, become lodged between floorboards, and subsequently circulated in the home by the forced-air furnace.

> The couple covered over the affected areas outside their home with bark chips, and Olin's mother wet-mopped three times a day. But eventually the family moved out. "I got tired of living in fear and I wanted another child. I didn't want to look back and regret that we hadn't left. Now we are staying with family members who, unfortunately, live where there are lots of old houses being renovated, so we're not without worries."

"My pediatrician says there isn't a lead problem in our state and my boys don't need to be tested. However, I live in an old home and I'm concerned. Should I go ahead and get them tested for lead?"

Absolutely. You may have to insist on it because requiring a widespread program for testing children for lead has

---

The amount of lead in the blood that the CDC considers acceptable has changed dramatically over the last fifty years. Here's how the number has trended down since 1960:

1960: 60 µg/dl

1970: 40 µg/dl

1975: 35 µg/dl

1985: 25 µg/dl

1991: 10 µg/dl

---

been hotly debated. In 1990 the CDC called for universal screening of all US children between the ages of one and five. But pediatricians complained that it was a cost burden for their middle-income patients and HMOs claimed that it was a money loser. In a climate of program cuts and unsupportive providers, the CDC stopped mandating universal testing and left it to individual states to determine their own requirements.

"Does socioeconomic status affect lead exposure?"

Yes. In poor cities children are more often exposed to lead. When the NHANES III studied blood lead levels in the US population from 1988 to 1991, they found that 35 percent of poor black children who lived in cities had blood lead levels equal to or greater than ten compared to 5 percent of white middle- and upper-income children living outside metropolitan areas. Poor children tend to live in older housing that is not well maintained, and even if lead is identified their families cannot afford to move out.

However, even though lead hits poor children dispro-

"Severe" lead poisoning is defined as a lead count of 45 µg/dl or higher.

"Moderate" lead poisoning is a lead count of 20–44 µg/dl.

Emergency blood replacement (called chelation therapy) can help prevent death in children whose lead level exceeds 45 µg/dl, but it does not eliminate neurological damage. Currently no treatment is prescribed for children who suffer from lower levels of lead exposure.

portionately, it does harm across the income spectrum. Any child living in a pre-1978 home, whether a tenement or a mansion, is at risk. The myth that lead is only a problem for low-income, inner-city children is dangerous in that it lulls other parents into thinking they do not have to worry about lead exposure and do not need to have their children tested for lead.

Real Stories

Harrison was born in Queens, New York, in 1997. His father is a sound engineer; his mother a stay-at-home mom. In those days, as now, apartments were enormously expensive in Manhattan, so the couple found a place to live in the more affordable neighborhood of Forest Hills, Queens, a mere fifteen-minute subway ride from midtown Manhattan. When they first moved in, the walls had recently been painted linen white and sun poured in through large-paned windows.

NORTHLAKE PUBLIC LIBRARY DIST.
231 N. WOLF ROAD
NORTHLAKE, IL 60164

At one year of age, as is law in New York State, their firstborn was screened for lead. When his blood lead level came back at fourteen, John and Jen feverishly tested the paint on the walls, the window frames, and their two antiques. They learned that the action of raising and lowering their windows during the hot summer months had incidentally abraded old paint, creating a residue of tiny paint chips and dust that settled inconspicuously on various surfaces. Harrison was exposed to lead by either breathing in minuscule amounts of lead when windows were opened and closed or, as he crawled through the apartment, he was getting specks of lead paint on his hands that he then put in his mouth. But Harrison—like other children who have had high lead levels at some point in their childhood—is turning out fine. Now in the fourth grade, Harrison does well in school and seems to have no impairment. His current teacher reports, "He is truly one of my best students—a very, very bright child. I would've never guessed he was exposed to lead at any capacity."

How do children like Harrison do it? The answer may lie with their parents. A recent study suggests that stimulating activities can help develop brains injured by lead.

Study: A stimulating environment minimizes the damaging effects of lead

To determine whether or not lead-exposed rat pups in enriched environments suffer fewer deficits than lead-exposed rat pups in settings that offer little stimulation, in 1997 Drs. Jay Schneider and Theodore Lidsky conducted a study. They split a population of rats into two groups and exposed them to distinct environments—one a cavelike cage and the other a habitat with wheels and tubes. They then fed them lead in their drinking water. The lead-exposed rats in the impoverished environment showed dramatic learning deficits. In the enriched environment, the lead-exposed rats performed similarly to their unexposed counterparts in a control group. This translates into practical brain-building activities, like reading, for all children, but especially those who have had some lead exposure.

Real Stories

Ralph Spezio is an energetic former school principal from Rochester, New York, who spent his career working with children from poor neighborhoods. When he suspected lead was keeping his students from their potential, he got access to the blood screening results for 112 of his preschool and kindergarten pupils. He learned that every single one of these students had a lead count greater than ten micrograms per deciliter.

"In my last year, before retiring, I told my staff we were retaining 30 percent of first and second graders. These were lead-poisoned kids and we needed to adjust the curriculum to meet their needs." Spezio was acting on his assumption that, just as stroke victims have shown, the brain is elastic and to some extent can overcome

damage, so can children with high lead counts. He describes the prescribed curriculum: "We took the most lead-poisoned kids and worked on preemerging reading skills all day long." At the end of the year, Spezio and his faculty tested these students and felt victorious when it became apparent that they would only need to hold back seven rather than thirty children.

WHAT YOU CAN DO TO PROTECT CHILDREN FROM
LEAD

Improved government regulations, including bans on lead in paint, gasoline, and water pipes, have made huge inroads in reducing lead poisoning. However, even today, primarily as a result of lead-based paint in old homes, children are injured by lead.

Here's what you can do to keep them safe from lead exposure:

- At the very least, know if there is lead in your house by using a home testing kit—the kind that you buy at your hardware store with a little pen that turns pink when you rub it on a leaded surface. If you want to get a truly accurate analysis of lead in an old home, talk to an environmental company listed in your local yellow pages under Environmental Products and Services. A thorough assessment usually costs about $400.

- In an old home, maintain all painted surfaces. Chipped or peeling paint can be hazardous.

- Renovations in a pre-1978 home should take place when no children or pregnant women are living there. A certified lead abatement professional should be responsible for cleaning up the site.

- If you live in an old house it is possible that dust from lead paint may be released whenever windows and doors are opened and closed. Clean sills near the window sashes and floors regularly with a damp cloth.

- Have your children tested for lead when they are six months old and again at one year. If their count is above two micrograms per deciliter, have them tested again at two years of age, which is the time lead levels generally peak. If they have had any known exposure, you may wish to continue annual testing until age six.

- Insist that your pediatrician give you the exact number of your child's blood lead level.

- If you have lead pipes, let cold water run for thirty to sixty seconds before using it for drinking or cooking. Never use hot tap water for food or food preparation. Hot water leaches more lead than cold.

- If you work in construction where you might be exposed to lead dust, remove your clothing before coming into your home or into contact with your children. Do not kiss or hug your child until you have bathed.

- Use a HEPA vacuum cleaner and HEPA air filter (products that do not recirculate dust) if there is any risk of lead particles in your home.

FOR FURTHER INFORMATION
www.healthychildrenproject.org/actions/lead.html
www.cdc.gov/nceh/lead
www.health.state.ny.us/environmental/lead
www.afhh.org/hah/hah_env_issue.htm
www.nlm.nih.gov/medlineplus/leadpoisoning.html
www.nsc.org/issues/lead
http://healthychild.org

3

MERCURY

FISH CONTAMINATED WITH MERCURY

Mercury is threat number two for children—an insidious and pervasive poison with especially bad effects on our children. Unlike lead, where contamination is on the decline, mercury is released into the air every day from coal-burning power plants. From the atmosphere it goes into water, where it ends up in fish, which concentrate quantities of methyl mercury in their flesh, depending on how high up they are on the food chain (fish that eat other fish—like tuna and swordfish—contain far more than, say, sardines). Of course, fish is also good for you. Because it is high in omega-3 fatty acids that promote brain development, it was not too long ago that pregnant women were advised to eat fish, particularly during the third trimester, when a baby's brain undergoes a growth spurt. For this reason health experts debate whether the benefits outweigh the risks. My research, however, leads me to believe pregnant women and young children should be *extremely* careful in choosing which kinds of fish to eat.

THE BAD NEWS

- A child is dangerously exposed to mercury in utero when a mother eats fish or seafood with a high level of mercury.

- Canned tuna and many kinds of sushi contain mercury.

- Prenatal exposure to mercury is much more powerful in causing intellectual and behavioral problems than exposure after a child is born.

- The CDC estimates that 16 percent of US women have blood mercury levels high enough to double their risk of giving birth to children with learning disabilities and/or neurological problems.

- No one knows for sure what level of mercury exposure in utero leads to demonstrable, lifelong harm.

- Mercury has been linked to:

- Memory deficits
- Shortened attention span
- Inability to concentrate
- Lack of coordination
- Problems learning language
- Poor vision and hearing
- Loss of IQ points
- Mental retardation
- Seizures
- Depression
- Bipolar disorder

THE GOOD NEWS

- You can keep your children safe from high levels of mercury exposure simply by avoiding fish altogether, or by referring to the list of low-mercury fish at the end of this chapter.

- Many studies have shown that ethyl mercury—used as an ingredient in many childhood vaccines until 2001—is not harmful, and in particular is not linked to the rise in autism.

- Today vaccines contain little to no ethyl mercury.

THE SCIENCE

The EPA and the FDA (US Food and Drug Administration) determined what they considered to be a safe amount of mercury in the bloodstream of pregnant women and children by analyzing studies of mercury poisoning in communities where the primary diet is fish. However, two studies that both agencies took into account in their safety calculations gave conflicting evidence about the toxicity of low-level doses of mercury.

The Faroe Islands study: Low levels of mercury exposure are toxic

People who live on the remote Faroe Islands in the North Atlantic Ocean, between Norway and Iceland, make their living fishing and processing meat. A typical Faroe Islands mother's diet primarily consists of seafood and whale meat. To better understand the effects of mercury in children, in 1987 Harvard University researchers traveled to the Faroe Islands and tested samples of umbilical cord blood and hair of 1,022 Faroe Islands children at birth, at age seven, and then again at age fourteen. Tests conducted when the children were seven showed that exposure to methyl mercury in the womb could be correlated with language deficits, an inability to concentrate, poor motor functions, and, to a lesser extent, poor visual-spatial perception. This study showed that the effects of methyl mercury exposure on brain function are detectable at levels currently considered safe.

The Seychelles Islands study: Low levels of mercury exposure are not toxic

To complicate things, the Seychelles Islands study contradicts the Faroe Islands study. Seychelles Islands mothers eat fish with levels of methyl mercury similar to those of fish sold in the United States; however, with an average of twelve fish meals a week, these women get a bigger dose of methyl mercury. Therefore, a Seychelles Islands baby has far greater exposure than a baby born in the United States. To determine whether these in utero exposures were harming Seychelles Islands babies, in 1995 Dr. Gary Myers and his colleagues from the University of Rochester tested 779 Seychelles Islands children at birth and then again at six months, nineteen months, twenty-nine months, and sixty-six months. What they determined is that low-level exposure to methyl mercury does not harm a baby's brain. According to Dr. Myers the current FDA standards are reasonable.

Although these two studies had very different results, the EPA advisories, as well as a prominent National Academy of Sciences report on mercury, evaluated all of the studies done on mercury exposure, including the abovementioned two, and realized there were additional variables in Myers's Seychelles Islands study. Therefore, in determining exposure levels, they placed much more emphasis on the Faroe Islands study as well as a study done in New Zealand, with support from adult studies conducted in the Amazon.

"How are children exposed to mercury?"

The CDC tells us that, by and large, the most common way children are exposed to methyl mercury is by either eating contaminated fish or being exposed in the womb when their moms eat the fish. Here's how it enters the environment and moves up the food chain:

Mercury occurs naturally in coal, so by far the largest source of methyl mercury in the United States is coal-burning power plants. However, purified mercury is also used commercially in fever thermometers, fluorescent lightbulbs, some electrical switches, and thermostats. Inevitably these mercury-containing items end up in junkyards and landfills, where the mercury vapor is released into the air. Believe it or not, crematoriums also release mercury vapor. When dental amalgam is incinerated, mercury vapor travels up the crematorium's smokestack into the environment.

Once mercury vapor is pumped into the air, it can travel on wind currents hundreds of miles from its source, raining down into lakes, rivers, and oceans. In our waterways, mercury is taken in by bacteria in aquatic sediment and tiny plants called plankton transform the heavy metal into methyl mercury. Next, small fish eat the plankton and bacteria and the methyl mercury magnifies in their tissues. Concentrating the poison with every link up the food chain, methyl mercury grows more potent as big fish swallow small fish. High up on the food chain, large fish like sharks and swordfish—as well as the birds that eat fish—have the highest concentration of methyl mercury. At the top of the food chain, humans are exposed to the highest dosage of methyl mercury when we eat contaminated fish. And at the tippy-top are our babies in utero who, in turn, receive an even larger concentration of mercury than their mothers, who complete the process of biomagnification by pumping blood across the placenta to the developing child.

Study: The social cost of mercury poisoning

Dr. Leo Trasande, a pediatrician at Mount Sinai's Center for Children and Environmental Health in New York City, is young and soft-spoken, with closely cropped black hair. During his residency in pediatrics at Harvard Medical School he began to see many children who were sick because of environmental factors. "I suddenly understood that we were facing a wide array of urgent epidemics as a result of environmental toxins." He realized that he would never address this epidemic one child at a time. To influence lawmakers, he decided to do the ultimate brass-tacks calculation for one particular toxin: to figure out the amount of money lost each year as a direct result of mercury contamination from coal-burning power plants and incinerator emissions. By adding up all the costs—including special education, health-care needs, and loss of workforce—due to deficits in brain power resulting from fetal exposure to mercury, he came to a figure of $8.7 billion a year.

"How long before I get pregnant do I need to think about the fish I am eating?"

Dr. Trasande suggests that women can protect their future unborn children by reducing their exposure to mercury six to twelve months before conception. However, we all know that pregnancies are not always planned in advance. One study done in 2001 revealed that 49 percent of pregnancies in the United States were unplanned. Whether your pregnancy is planned or not, follow the fish-eating guidelines outlined at the end of this chapter.

"How many women in this country have mercury levels high enough to harm their unborn babies?"

By analyzing studies of communities poisoned by mercury, the EPA has determined that mercury above 5.8 parts per billion in the bloodstream of a woman of childbearing age can harm babies in the womb. However, if a pregnant woman eats fish with a high mercury concentration, the metal is pumped across to the fetus and becomes concentrated in the baby's umbilical cord blood, significantly upping the dose of mercury in fetal blood above the mother's own mercury blood level. For this reason, many

Q & A:
Dr. Leo Trasande

Q: Is the EPA's threshold level for methyl mercury safe for our babies in utero?

Dr. Transande: There is increasing evidence that the EPA's safety level for methyl mercury is not sufficient.

researchers believe the EPA's threshold may not be high enough to protect developing babies in utero.

The EPA's questionable mercury threshold is reminiscent of the CDC's disputed lead threshold. In the 1960s the CDC determined that anything under sixty micrograms per deciliter of lead kept children out of harm's way. Researchers now know that even a count of two effectively hammers away at the developing brain, shaving off IQ points. While researchers have been examining lead toxicity for a hundred years, mercury has only been studied for fifty. Only time will tell whether the EPA's current methyl mercury threshold level is appropriate.

"Are learning disabilities linked to mercury exposure?"
The National Academy of Sciences estimates that mercury exposure in the womb is responsible for sixty thousand babies born in this country each year with neurological

damage and mental impairments. The Learning Disabilities Association of America (LDA), the National Education Association (NEA), and the Arc of the United States all say mercury pollution is one of the biggest culprits in the rising tide of learning deficits. However, one popular argument for the dramatic increase in learning disabilities is that we are more readily diagnosing children. This viewpoint suggests that back in the 1970s and '80s children who had trouble concentrating and learning to read were simply mainstreamed and had to learn to adapt; or perhaps they lost patience and dropped out.

Interested by this question, I talked to three women who have taught learning-disabled children for more than twenty years. I wanted to get their sense of whether the climbing numbers can be attributed to more frequent diagnoses or whether they truly represent a dramatic increase in children who have trouble learning and processing information. Liz Bosworth, Collette Bonelli, and Joan McGowan all teach in a school district that encompasses schools in and around Plattsburgh, New York. All three women are intelligent and, chuckling as they complete one another's thoughts, their friendship runs deep. "You've got to be a team when you get inside those classrooms," explained Collette, "and that makes you close."

When I asked them about autism, Joan, a striking woman with thick red hair, blurted, "Oh, yeah, we're alarmed. It's constant—the autistic kids." She went on to explain that when she first began to teach special education more than two decades ago she had one autistic student, and today the number of children on the autistic spectrum is mushrooming.

Although they agreed that they believe more children with learning disabilities are showing up in their classrooms, they felt it was because of the competitive nature of the federal program No Child Left Behind. Evidently classroom teachers do not want children who are mildly impaired bringing down their test scores. The women did, however, express concern about a significant increase in students' poor coordination and behavioral problems. "I'm seeing far worse motor skills—both fine and gross," mentioned Liz, a tiny, muscular woman with an infectious smile. "It's nearly impossible for some of them to hold a pencil"—she curled her fingers in on themselves—"almost like they have cerebral palsy."

Real Stories

It is only natural for mothers of children struggling with learning disabilities to speculate that their child's disability might be the result of mercury in tuna salad sandwiches they toted to work in lunch bags, sashimi gobbled down at sushi bars, or salmon steaks tossed on the grill. To learn if there is a connection between the mercury in the fish she ate while she was carrying her son and the boy's impaired motor skills—even at age nine, he has difficulty forming letters and tying his shoes—one mom sent a treasured lock of her son's baby hair to a

> *lab in Wisconsin. A chemist with the state department of public health tested the thin wisps of baby hair. The ultrafine blond tips of hair revealed that this boy's mercury level was 0.617 micrograms per gram (parts per million). This mom was relieved to learn that although her son's mercury level at birth was consistent with someone who ate moderate amounts of fish, it measured well below the current EPA/FDA threshold of 5.8 parts per billion.*

"Is there still mercury in vaccines and is it responsible for the increase in autism?"

From the 1930s until 2001 a preservative called thimerosal, which kept vaccines from becoming contaminated, was used in shots given to children. The preservative is 49.6 percent mercury by weight and is metabolized into ethyl mercury that is distinct from methyl mercury found in fish. Because ethyl mercury stays in the blood for a shorter period of time than methyl mercury—its half-life is a week or two—many researchers believe that it is eliminated by the time the child receives his next vaccine.

Numerous studies have looked for—but not found—a link between mercury in vaccines and autism. For example, the CDC and the National Institutes of Health asked the Institute of Medicine, an arm of the National Academy of Science, to review the safety of child immunizations, especially the link between the dramatic rise in autism and the measles, mumps, rubella (MMR) vaccine and thimerosal. The Institute of Medicine's 214-page report from May 2004 found no connection between thimerosal and autism: "The hypotheses regarding a link between autism and thimerosal-containing vaccines lack supporting evidence and are only theoretical."

Since then, studies in countries where thimerosal was removed from vaccines sooner than in the United States have not shown a decline in autism. Despite those studies, and in part because of limitations in their design and interpretation, some scientists and advocacy organizations continue to search for a link. One possibility currently being investigated involves the potential for mercury to be toxic to the immune system in some susceptible children and thereby increase the risk of autism. This hypothesis is under investigation by a number of scientists.

By 2001, the Public Health Service Agencies (PHS), the American Academy of Pediatrics (AAP), and vaccine manufacturers had agreed to remove thimerosal from most vaccines given to children.

The FDA released a report in 2006 that stated that the Diptheria, Tetanus, Pertussis (DTaP) shot manufactured by Tripedia (Sanofi Pasteur, Inc.) and some flu shots are the only immunizations routinely given to children under six that still contain thimerosal. To be safe, it is important to ask your pediatrician if your child's immunizations contain thimerosal. If so, you can request that your child be given the vaccine from a manufacturer that does not use thimerosal in their products.

"Is there still mercury in flu shots? I was under the impression that mercury had been taken out of all vaccines for children."

Yes. Thimerosal is still used in some flu shots. Because of the lingering controversy over ethyl mercury and children's health, there is no harm in practicing the precautionary principle by asking for a thimerosal-free flu vaccine for your children as well as for yourself if you are a woman of childbearing age.

"Do I need to worry about the mercury in dental fillings?"

Used for the last hundred years, dental amalgam is 50 percent mercury, as well as a combination of other metals, including silver, copper, tin, and zinc. The amalgam used to fill teeth is initially putty. Pushed down into a tooth, it takes the shape of the cavity. Amalgam has been a favorite of the American Dental Association because it is easy to use, cheap, and it doesn't wear out. Any silver-colored fillings your children have in their mouths are probably amalgam. The World Health Organization says that amalgam fillings release about one microgram, or one-millionth of a gram, of mercury into your body every day. Comparatively, the EPA says that each of us takes in five to six micrograms of mercury in our food and water daily. Although many European countries have restricted the use of amalgam fillings, there is no scientific proof that they are harmful for children.

Study: Amalgam fillings and our children's health

From 1997 to 2005 Dr. David Bellinger from Harvard Medical School's Department of Neurology conducted a study in Boston and rural Maine that examined 534 children who had unmet dental needs, and therefore had no amalgam in their mouths. He administered a series of tests to measure IQ, neural functions, behavior, and kidney functions. Then he had the cavities of half the children filled with amalgam and those of the other half filled with composite resin material. The children were tested yearly and then at the end of five years. Although the amalgam group had 50 percent higher mercury in their urine, there appeared to be no correlation between level of methyl mercury exposure from dental amalgam and IQ, neural functions, behavior, or kidney functions.

Dr. Bellinger remarks that one issue with the study is that it did not include younger children. Mercury is most dangerous for children before the age of six, when the nervous system is fully developed. On the other hand, relatively few children get dental fillings at such an early age.

"What are the alternatives to silver dental fillings?"

Fifteen years ago more than half of dental fillings were made from amalgam—a figure that has dropped to 30 percent today. Instead most cavities are now filled with resin composites. White to match teeth, they are made of powdered glass and epoxylike materials. Because composites are not as durable as amalgam, many dentists still prefer amalgam for large cavities in back teeth. Composites cost about 50 percent more than amalgam fillings.

I recommend asking what materials your child's dentist uses to fill cavities and requesting that he or she not use amalgam. Although Dr. Bellinger's study found no measurable effects from the amount of mercury contamination attributable to amalgam fillings alone, he says, "It's a crazy idea to put a heavy metal in a child's mouth."

"Should I have my child's amalgam fillings removed?"

Unless there is a problem, most experts say no. Not only is it expensive (and not covered by insurance), but the removal process can expose your child to a spike in additional mercury exposure. It's worth noting, however, that although there is no scientific proof that removing amalgam fillings improves health, there is anecdotal evidence that it has improved a number of health conditions like chronic fatigue syndrome and multiple sclerosis.

"MERCURY THREAT TO KIDS RISING, UNRELEASED
EPA REPORT WARNS"
—*Wall Street Journal* (February 20, 2003)

A single high mercury exposure can harm a fetus at a critical time of brain development.

GUIDELINES FOR EATING FISH

In 2004 the EPA and FDA issued this advice for women who might become pregnant, women who are pregnant, nursing mothers, and young children:

> Some fish and shellfish contain higher levels of mercury that may harm an unborn baby or young child's developing nervous system. The risks from mercury in fish and shellfish depend on the amount of fish and shellfish eaten and the levels of mercury in the fish and shellfish. Therefore, the Food and Drug Administration (FDA) and the Environmental Protection Agency (EPA) are advising women who may become pregnant, pregnant women, nursing mothers, and young children to avoid some types of fish and eat fish and shellfish that are lower in mercury.

The Environmental Working Group (EWG), an environmental advocacy group out of Washington, DC, and *The Green Guide*, a monthly on-line publication, have both put out guidelines that are slightly more stringent than the FDA and EPA's. Because mercury is so toxic to the developing brain it is wise to err on the side of precaution. Therefore, pregnant women, nursing mothers, women who may become pregnant, and young children should either

---

There is no way to prepare or cook fish so that it is safe from mercury contamination.

---

eliminate fish altogether or limit low-mercury fish to one serving per week, moderate-mercury fish to one meal per month, and avoid high-mercury fish entirely.

- **High-mercury fish.** Stay away from species with a high-mercury content (especially if you are pregnant): Atlantic halibut, king mackerel, marlin, oysters (Gulf Coast), pike, sea bass, shark, swordfish, tilefish (golden snapper), tuna (steaks and canned albacore).

- **Moderate-mercury fish.** Once a month you can eat moderately contaminated fish: Alaskan halibut, black cod, blue crab (Gulf Coast), blue mussels, cod, Dungeness crab, Eastern oysters, mahi-mahi, pollack, tuna (canned light).

- **Low-mercury fish.** Once a week you can eat fish with low-mercury levels: anchovies, arctic char, blue crab (mid-Atlantic), clams, crawfish, croaker, farmed catfish,* farmed trout,* flounder, haddock, herring, king crab, Pacific sole, Pacific salmon, sand dabs, scallops, shrimp, striped bass, sturgeon, tilapia.

*It is important to remember that although some farmed fish have low mercury levels, they may contain PCBs that can harm babies in utero.

"MERCURY AND TUNA: U.S. ADVICE LEAVES
LOTS OF QUESTIONS"
—*Wall Street Journal* (August 1, 2005)

"TUNA ON RYE—HOLD THE MERCURY, PLEASE"
—*O, The Oprah Magazine* (April 2005)

"Is canned tuna safe?"

The FDA claims that eating one ounce of canned tuna for every twenty pounds of a person's body weight a week is safe. Seeing that a regular can of tuna is six ounces, a sixty-pound child can safely eat about half a can of tuna a week. However, as noted above, the EWG and *The Green Guide* have issued a more careful assessment that suggests that canned light tuna has moderate amounts of mercury and should only be eaten by women of childbearing age and small children once a month.

"What about fish sticks and fish sandwiches?"

The EPA assures us that fish sticks and fish sandwiches are made from fish low down on the food chain, meaning they are low-mercury fish.

*"My husband and son like to go fishing in the river
behind our house. Is it safe for them to eat the fish?"*

It most likely is not safe for them to eat fish from local lakes, streams, or ponds. By 2004 the EPA reported that forty-eight states—all except Alaska and Wyoming—had issued fish consumption advisories due to mercury contamination in their lakes and rivers. In fifteen states this applies to 100 percent of their waterways. Additionally, 92 percent of the eastern seaboard and 100 percent of the Gulf Coast is under advisory.

**STATES WITH FISH CONTAMINATION ADVISORIES
IN 100 PERCENT OF THEIR WATERWAYS**

Connecticut	New Hampshire
Florida	New Jersey
Illinois	North Dakota
Kentucky	Ohio
Maryland	Pennsylvania
Massachusetts	Rhode Island
Minnesota	Vermont
Missouri	

The EPA suggests that before you go fishing you should take a look at your state's fishing regulations booklet that you receive when applying for your fishing license. It will have information about mercury contamination in local bodies of water. You can also get information about local fishing advisories by contacting your local health department.

WHAT YOU CAN DO TO PROTECT CHILDREN FROM
MERCURY

- Make smart choices by consulting the list (above) of mercury levels in various fish before feeding it to your children or eating it yourself if you are pregnant, nursing, or intend to become pregnant in the next year.

- To calculate acceptable levels of mercury consumption—which vary by weight of the consumer—go to www.gotmercury.org.

- If you are thinking of becoming pregnant, test your own level of mercury. You can buy a mercury test kit for $25 at http://usa.greenpeace.org/mercury.

- Call or e-mail your senators (www.senate.gov) and representatives (www.house.gov/writerep) and tell them you don't want mercury in your fish or in your body. Urge them to support legislation that would reduce the acceptable level of mercury emissions from coal-burning power plants and chemical factories. Mercury emitted from these sources ends up in bodies of water and contaminates the fish we need to eat for good health.

FOR FURTHER INFORMATION

FDA mercury advisory: www.cfsan.fda.gov/dms/adme hg3.html

FDA food information (toll-free): 1-888-SAFEFOOD

EPA fish advisory: www.epa.gov/waterscience/fish

EPA mercury information: www.epa.gov/mercury

Environmental Working Group: www.ewg.org

The Green Guide: www.thegreenguide.com

4
PLASTICS

TOSS THE BAD PLASTICS

Unlike lead and mercury, the dangers of plastics are not widely understood by the public. There are two villains here: a family of chemical compounds called phthalates that make PVC plastic soft (as in plastic shower curtains or wash-off baby books that go in bathtubs) and the industrial chemical bisphenol A (a hard, shiny plastic often used for baby bottles and sippy cups). Although the chemicals involved are different they have a similar effect, mimicking hormones. As they leach out of the plastic they *feminize* babies and small children. This leads to a variety of problems, including smaller penises as well as planting the seeds for cancer that can develop later in life. The plastics industry may squeal about this one, but I *strongly* urge you to remove from your home any plastics you suspect might contain these chemicals. Because they are often hard to identify, I have tried to replace most plastic items in my home with glass, wood, and stainless-steel alternatives.

THE BAD NEWS

- Scientist call these feminizing chemicals endocrine disruptors. Endocrine disruptors interfere with the normal functioning of hormones, including estrogen and thyroid.

- Exposure to these hormone-mimicking chemicals during critical periods in a child's development, both in utero and in infancy, can result in lifelong injury.

- Estrogen is one of the major communicators in both the male and the female body, telling cells how they should behave. When fetal cells are exposed to even a tiny amount of additional estrogen, the reproductive system can become reprogrammed, resulting in early onset adolescence, undescended testicles, enlarged prostates, reduced sperm counts, and smaller penises.

- The reproductive abnormalities associated with these chemicals set the stage for cancers later in life, such as breast cancer or prostate cancer.

- If you have these plastics in your home, their harmful chemicals spread everywhere, even accumulating in dust bunnies.

■ In rodent studies, phthalates are linked to:
  • Liver cancer
  • Damaged kidneys
  • Slightly smaller scrotums
  • Undescended testicles
  • Smaller penises
  • Hypospadias (a birth defect where the opening of the urethra is on the base of the penis rather than the tip)
  • Reduced sperm count
  • Reduced testosterone
  • Testicular cancer

THE GOOD NEWS

■ This chapter specifically outlines which plastics to avoid.

■ It is easy to identify the offending plastics and keep them out of your home.

■ Many countries—though not yet the United States—have prohibited the use of endocrine disruptors in consumer products, especially those marketed for children. Once banned, levels of these chemicals have quickly dropped in people's bodies.

THE SCIENCE

PHTHALATES

Also referred to as plasticizers, phthalates are a family of eight hormone-mimicking chemical compounds. These compounds are the additives that make plastic soft and flexible in everything from toys, shower curtains, and plastic bags to vinyl floors and medical tubing. They are also used to enhance personal-care products by making perfume last longer, nail polish more elastic, and lotions more easily absorbed into the skin.

"STUDY FINDS GENITAL ABNORMALITIES IN BOYS: Widely used industrial compounds . . . linked by researchers to changes in the reproductive organs of male infants."
—*Los Angeles Times* (May 27, 2005)

Study: Phthalates feminize baby boys in utero

If you want to be able to spot phthalates think soft plastic, like wash-off baby books, plastic shower curtains, medical tubing, kitchen flooring, food packaging, and even some toys. Hard plastic cars that children pedal around the backyard do not contain phthalates.

Q&A: Dr. Shanna Swan

Q: Should couples reduce phthalate exposure before becoming pregnant?

Dr. Swan: Pregnant women or couples attempting to conceive may want to limit their use of phthalate-containing personal-care products.

In June 2005 the first-ever published study on in utero phthalate exposure linked this exposure to abnormalities in baby boys' genitalia. Dr. Shanna Swan from the University of Rochester found this connection by measuring phthalate monoester metabolites in the urine of eighty-five pregnant women in Los Angeles, Minneapolis, and Columbia, Missouri, and then later examining their sons soon after birth. She and her team found that mothers with the highest level of phthalate metabolites in their urine late in pregnancy had baby boys with smaller penises and scrotums, incomplete descent of the testicles, and a shorter perineum—which scientists call anogenital distance (AGD)—about the same length as on normal females. (Anogential shortening was also found in rodent studies the EPA conducted.) Given that normal males have a lon-

About 25 percent of US women have levels of phthalates high enough to affect the genital development of baby boys in the womb.

ger AGD than females, Dr. Swan considers a shortened AGD a marker for demasculinization.

Although it is not clear to Dr. Swan and her team that long-term repercussions result from these idiosyncrasies, in three to five years she plans to reassess the eighty-five boys analyzed. In particular, she will be looking at play activity to understand if boys with the highest level of phthalates display typically "feminine behaviors." She explains, "For the initial look we will ask moms to report on the toys their children play with, as well as the games they most frequently play." She goes on to say, "There are some well-validated scales that we can use to rate these responses on a male-typical versus female-typical spectrum."

Study: Phthalates in rodent studies by EPA

In rodent studies the EPA conducted, rats exposed to phthalates contracted liver cancer and developed damaged kidneys; in addition, male rat pups exposed to the chemicals in utero experienced dramatic reproductive mutation, including smaller scrotums, undescended testicles, hypospadias, and reduced penis size. This cluster of abnormalities frequently results in lower sperm counts, infertility, reduced testosterone, and testicular cancer, a phenomenon scientists now refer to as "phthalate syndrome."

"Are some phthalates more dangerous than others?"
There are eight chemical compounds in the phthalate family, some more toxic than others. The three most potent phthalates are diethyl phthalate (DEHP), dibutyl phthalate (DBP), and benzylbutyl phthalate (BBP). Not only do they adversely affect health by themselves, but even in small doses they interact with one another in ways we don't understand. The dominant phthalate, DEHP, which is in, among other things, shower curtains, cable sheathing, garden hoses, and some toys, has been used so widely that it can now be found literally all over the world: in subsurface snow in Antarctica and in jellyfish more than three hundred feet below the surface of the Atlantic.

WHAT PRODUCTS CONTAIN THE BIG THREE?

- DEHP: vinyl products, floor tiles, upholstery, shower curtains, cables, garden hoses, rainwear, car parts and interiors, packaging film, sheathing for wire and cable, some food containers, toys, and medical devices

- DBP: nail polish, cosmetics, and insecticides

- BBP: adhesives, paints, sealants, car-care products, vinyl flooring, and some personal-care products

Real Stories

Environmental websites frequently report a link between phthalate exposure in the womb and a birth defect called hypospadias, a condition in which the urethral opening is not at the tip of the penis, but instead, in the most severe cases, at the base, or in the least severe, slightly off center. The CDC's website says that the incidence of hypospadias increased in the 1970s and 1980s, when phthalates began to be widely employed in consumer products. The incidence of this defect, however, began to stabilize in the 1990s at about thirty to forty cases per ten thousand births.

The Organic Consumers Association published an article in June 2004 that mentions Olivia James, a former fashion model who believes that chemicals in the makeup she used while modeling were responsible for her son's hypospadias. A thoughtful, professional woman who now works as an analyst at Dow Jones, Olivia described to me the heavy applications of makeup she wore frequently during the fifteen years she modeled. She spoke about the difficulties surrounding her son's birth three years after she retired from the fashion business. Finally, she described the article she read when her son was three, reporting studies linking phthalates used in cosmetics to an increased incidence of hypospadias in lab animals. She remarked, "I felt so guilty. That maybe things would have been different if I hadn't chosen to be a model." That report, coupled with the CDC's findings, led Olivia to draw the connection between her use of heavy cosmetics until 1993 and her son's birth defect in 1996. Although phthalates cross the placental barrier and are passed to our babies in our breast milk, they are not persistent like PCBs and flame retardants. Instead, even though our moment-to-moment phthalate body burden appears to be constant because of continual exposure, they quickly pass through our systems. Although Olivia was not modeling at the time she was pregnant she continued to use beauty products. "I didn't wear makeup, but I did use hair gel, hair spray, and a hair straightener once a week."

I talked to Dr. Ted Schettler, science director of the Science and Environmental Health Network, about this suspected link. He told me, "Some scientists think that antiandrogenic phthalates like DEHP, DBP, and BBP could contribute to hypospadias through the reduction of testosterone levels in the male fetus. And it is true that beauty products often do contain DBP, and sometimes even other phthalates. It's unclear, however, how much cosmetics actually contribute to DBP exposures. So this remains only an hypotheses."

"How are our children exposed to phthalates?"

Whether we like it or not, plasticizers from our computer cables, the upholstery on our sofas, and even some of the toys strewn across playroom floors nationwide end up inside our children. It is probably fair to say that every child in America is born having had exposure to phthalates in the womb.

Manufacturers are not required to list phthalates as an ingredient on product labels, making it nearly impossible to limit contact with these hormone-mimicking compounds. As a result, like everyone else, as our teenagers work gel into their hair, swipe antiperspirant under their arms, breathe common household dust, and eat foods packaged in plastic wrap, they get a steady dose of plasticizers. Because these chemical compounds are not stable, when small children put a soft plastic toy, for instance a rubber duck, in their mouth they are treated to a dose of phthalate. Furthermore, if they ever need a medical treatment, such as a blood transfusion or an intravenous feeding tube, phthalates would leach directly into their veins.

THE CDC REPORTS THAT CHILDREN ARE EXPOSED TO PHTHALATES BY:

- Breathing air contaminated with phthalates that have migrated out of phthalate-containing products in our homes, like shower curtains, vinyl flooring, garden hoses, and some plastic toys.

- Mouthing soft plastic toys that contain phthalates.

- Applying personal-care products that contain phthalates, especially to highly absorbent areas of the body, like armpits, the palms of the hand, or the scalp.

- Eating food that has come into contact with packaging that contains phthalates and bisphenol A.

- Receiving a medical treatment like a blood transfusion or IV feeding tube that uses medical tubing that contains phthalates.

Study: Sick babies get phthalates in intensive care units

As it turns out, a study conducted by Dr. Howard Hu at the Harvard School of Public Health shows that newborn babies in hospitals receive two to three times more than

Q&A: Dr. Howard Hu

Q: Is there anything parents can do to reduce their children's phthalate exposure in neonatal intensive care units?

Dr. Hu: Parents should make a point of asking the neonatal staff if they are using low-phthalate products. Such products are definitely available. Babies with serious health problems require IV bags, blood bags, and medical tubing. Phthalates are the least-expensive alternative for making these devices flexible.

the average daily DEHP dosage the FDA considers safe for adults. During March and April 2003, researchers measured DEHP residue in urine taken from fifty-four boys and girls at two Boston-area neonatal intensive care units (NICUs). They learned that the more DEHP-containing medical devices used to treat a baby, the more DEHP entered his or her body.

The FDA issued a public health warning in 2002, recommending that pubescent boys, baby boys, and pregnant women carrying boy babies avoid medical devices that contain DEHP. However, when asked if hospitals were aware of the public health warning, the FDA toxicologist responsible for the 2001 safety assessment, said, "I don't know if this is on their radar screen." When asked if hospitals were implementing low-phthalate products in their neonatal intensive care units, he said, "Some hospitals are following the public health warning."

Lynn Spilman, an energetic clinical nurse specialist at the Albany Medical Center NICU, told me that in her unit, "We have been using low-DEHP-release products since 2001." At another major New York hospital, the nurse manager of the NICU I spoke with seemed confused about the public health warning and wanted to know what devices contained phthalates. She went on to say that her staff was never made aware of the FDA's public health warning and therefore had not replaced their NICU medical devices with the readily available, but somewhat more expensive, low-phthalate alternatives.

"To what degree are phthalates finding their way inside my children's bodies?"

The CDC has done studies to understand how much of this stuff is inside us. In 2001–02 its researchers analyzed phthalate residue in sample groups of children and women. It is unnerving to know that these studies found that children excrete high concentrations of phthalates in their urine, and that levels of dibutyl phthalate (DBP), which is found in nail polish and cosmetics, were particularly high for women of childbearing years.

"My daughter is starting to wear makeup. Do I need to worry about chemicals in her cosmetics?"

The CDC's findings regarding phthalates and women of childbearing years suggests that one of the chief delivery mechanisms for phthalates might be cosmetics, hair products, and lotions. A 2004 study conducted by the FDA found that two-thirds of the beauty products analyzed contained DBP. Still we don't know the health risks resulting from our preteens' and teens' growing affection for and use of makeup and hair care products.

> "SHOULD YOU WORRY ABOUT CHEMICALS IN
> YOUR MAKEUP?"
> —*New York Times* (July 7, 2005)

Although the beauty product industry trade group, the Cosmetics, Toiletries and Fragrance Association (CTFA), says phthalates are "safe as currently used"—the FDA has left it to the industry trade group to test products for safety—one wonders whether anyone has taken into account the heightened aggregate exposure to teenage girls who, in addition to common phthalate exposure, use numerous beauty products each day. Marian Stanley of the American Chemistry Council, another trade group, states her group's position: "There is no reliable evidence that phthalates used in cosmetics have caused any health problems in humans." Further, she maintains, "Primates, including humans, don't show a sensitivity to phthalates the same way rodents do."

As parents, it is difficult not to question Ms. Stanley's statements; it is also difficult not to wonder why we in the United States keep manufacturing plasticizers that have been shown to be harmful in animal studies. It is much easier to feel aligned with Dr. Earl Gray from the EPA, who states, "Rodent studies work quite well for testing androgen and antiandrogen activity. There is no evidence that rodent models are not appropriate."

Real Stories

To find out how a teenager's makeup, shampoo, and deodorant might affect her health, my seventeen-year-old, Paulina, and her two close friends, Ariel and Emily, decided to test the products they use on their hair and skin for harmful chemicals. They started by keeping a log of every personal-care item they used over a period of four weeks. Their enthusiasm inspired detailed inven-

tories that included even the minutest dab of lip gloss. Then, they sent their lists off to be assessed by Bryony Schwan, founder of Women's Voices for the Earth, the leading women's environmental advocacy organization.

Schwan submitted the girls' lists to the Campaign for Safe Cosmetics—an effort launched by a consortium of environmental groups in 2003 to inspire toxic-free beauty products and intelligent laws that keep harmful chemicals out of personal-care products. Their endeavor includes keeping a database that ranks twelve thousand brand-name beauty products by level of toxicity. (The database can be found at the Environmental Working Group website—www.ewg.org.) A month after submitting their lists of beauty products, Schwan sent Paulina, Ariel, and Emily an e-mail reporting that she had completed their beauty product assessments by running a total of twenty-one products through the Campaign for Safe Cosmetics database.

The analysis rated products from 0 to 10, with 10 being of highest health concern. Paulina, Ariel, and Emily each scored around 6.5 for their collective products. Schwan explained to them, "You've got two issues: these products either contain cancer-causing chemicals or they have ingredients which have not yet been assessed, so, as a consumer, you are left in the dark."

When pressed about phthalates, Schwan told me, "There were quite a few fragrances on their product lists that are protected as trade secrets. So although fragrances often contain DEP and DBP, there's no way for us to know for sure which ones do not." Finally, she said that if they were her daughters, she "would suggest they use as few products as possible, then I would try to get them to replace cosmetics that are either carcinogenic or linked to birth defects with others that are at the bottom of the scale."

It is comforting to know that 150 cosmetic companies, including the Body Shop and Kiss My Face, have recently committed to replacing all toxic chemicals in their products with safer alternatives. It is also encouraging that at least some governments have taken action to protect consumers from phthalates; for example, the countries of the European Union have recently adopted new cosmetic safety standards sharply limiting the use of phthalates and other harmful compounds. International cosmetic companies like Revlon and L'Oréal are reformulating products to meet the new standards. It is also a relief to learn that new legislation in California requires cosmetic companies to list phthalates on their product labels. Perhaps other states will follow their lead.

"Are there phthalates in the dust accumulating underneath my refrigerator?"

I am now speculating that phthalates have migrated out of the plastic ducks lined up along the edge of my bathtub as well as out of the vinyl tiles on my powder room floor and are indifferently swirling around in increasingly large dust

PERCENTAGE OF PHTHALATES IN HOUSEHOLD DUST

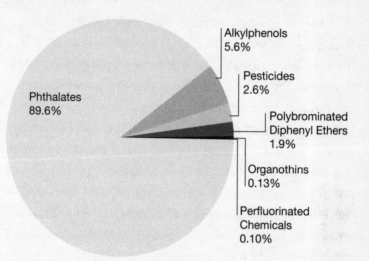

Figure 1
Average contribution of each group of chemical contaminants in the total concentration of all chemicals tested for in seven composite house dust samples

Note: This graph represents only the contributions of the six categories of chemicals tested in this study to the sum total concentration of all forty-four chemicals detected. The percents are not an indication of content in total dust quantity nor of all chemicals potentially present in house dust.

Reprinted with permission from CPA.

bunnies that take up residence underneath my children's beds.

To get a sense of what chemicals tend to accumulate in homes across America, I took a look at the Clean Production Action's (CPA) study of dust samples. In 2004, CPA—an environmental group that works with companies to replace toxic chemicals in their products with safer alternatives—gathered dust from vacuum bags in seventy homes nationwide. They then analyzed the dust samples for chemicals that are known to be harmful to the immune system and reproductive organs in animal tests. They discovered that phthalates made up 89 percent of the total concentration of chemicals found, and that DEHP alone comprised 69 percent of the forty-four toxins.

BISPHENOL A

Although the effects of bisphenol A (BPA) are not understood as well as those of phthalates, there is enough evidence to raise an alarm. Some countries, like Japan, have already banned its use.

Study: Bisphenol A in rodents

Professor Vom Saal at the University of Missouri has been researching endocrine disruptors since 1991 and looking at bisphenol A in particular since 1995. Although there are no human studies, in animal studies he has learned that small doses of bisphenol A injure reproductive organs in male pups, increasing the size of the prostate and decreasing sperm production. In the female pups exposed in the womb, Professor Vom Saal found that the estrogenic effects of bisphenol A alter uterine and breast cells, significantly increase body weight, and result in the early onset of puberty. The dosage given to mice in these studies is comparable to the amount of bisphenol A children are exposed to on a daily basis in this country.

Professor Vom Saal's research also links low-dose exposure to an array of behavioral outcomes in animals such as ADHD, hyperactivity, poor motor skills, and learning disabilities. How do researchers know if bisphenol A has the same effects in humans? Dr. Vom Saal explains, "We don't *know* for sure, but we see these growing health trends, starting in childhood—hyperaggressiveness, learning problems, and difficulty with social interactions and play behavior. And, later, if not in childhood, chronic anxiety. Some of these trends are so prevalent they almost seem normal: abnormal puberty changes, fertility difficulties for both men and women, breast cancer, prostate cancer. All of these trends parallel the onset of the plastics revolution that began twenty-five to thirty years ago. Part of this is just connecting the dots."

In animal studies bisphenol A is linked to:

- Early onset of puberty
- Obesity
- Larger than normal prostate
- Smaller sperm-carrying ducts
- Reduced sperm count

Bisphenol A can easily pass through the placenta, where absorption and distribution in the
fetus is rapid.

- Breast changes that represent early stages of breast cancer
- Altered immune function

"How does bisphenol A get into our children's bodies?"
Bisphenol A is used in polycarbonate, the hard, shiny plastic used to make many products, including clear plastic
baby bottles, toddler sippy cups, dental sealants, the interior coating of some food cans, sport bottles like Nalgene,
and five-gallon water jugs. The greatest exposure comes
not from touching these things, but when food and bev-
erages come in contact with bisphenol A, soaking it up,
especially when food and liquid are heated up in a polycarbonate receptacle. The more you heat the container, especially when it's scratched, the more bisphenol A you drive

First devised as a sex hormone drug in 1936,
chemists realized bisphenol A could be used to
make polycarbonate in 1962.

into the food. For this reason, never pour hot liquid into polycarbonate bottles, and if they are old or scratched, bisphenol A can migrate more readily into your food so it is best to throw it out. Because bisphenol A moves easily through the placenta, moms can pass it to their unborn babies.

"What can I do right now to reduce my family's exposure to bisphenol A?"

It's not easy, but do what you can to remove products with bisphenol A from your home. Based on what we have seen in countries like Japan, where they have banned bisphenol A, if you stop using polycarbonate, your children will experience a rapid drop in bisphenol A levels in their bodies.

Today, with more than 6.6 billion pounds produced yearly, clear plastic—made of polycarbonates—is everywhere.

"So where are these toxins coming from and are there safer alternatives?"

It is easy to fret about the possibility of infinitesimal amounts of hormone-mimicking chemicals gathering momentum as they band together in our children's bodies. Dr. Earl Gray, the EPA's chief phthalate researcher, underlines the concern by saying, "It's clear that chemicals that disrupt the role of hormones will, in small doses, act together and become ever more potent." Although there is no way to entirely elimi- nate these chemicals from our children's lives, it is possible to effectively ban most of the bad plastics from our homes and replace them with safer alternatives.

WHAT YOU CAN DO TO PROTECT CHILDREN FROM HARMFUL CHEMICALS IN
PLASTIC

Although the jury is still out on whether or not low-level exposure to hormone-mimicking chemicals harms our children's health, it is wise to take a few precautionary steps. Choosing to adopt even just a few of the following suggestions puts you ahead, reducing the level of exposure to your child.

Toys, Baby Bottles, Pacifiers, and Teething Rings
In the 1980s the EPA and the National Toxics Program (NTP) conducted studies that showed DEHP caused cancer in rodents. In reaction, the US Consumer Product Safety Commission (CPSC) asked toy makers to remove DEHP from any products that babies might put in their mouths, for instance rattles, teething rings, and pacifiers.

The European Union has gone further. Although phthalates are the cheapest option for softening plastic, these countries have recently mandated that any DEHP, BBP, and DBP in toys sold in any of its twenty-five member countries be replaced with nontoxic alternatives. The EU has prohibited three addtional suspicious chemicals—diisononyl phthalate (DINP), diisodecyl phthalate (DIDP), and dioctyl phthalate (DOP)—from pacifiers and baby bottle nipples, as well as other articles for children designed to go in the mouth. This represents a progressive governmental action affecting twenty-seven countries (as of 2007), benefiting the health of their 457 million inhabitants.

BABY BOTTLES. Many baby bottles are made of polycarbonates that contain bisphenol A. Unfortunately, manufacturers are not required to label baby bottles. Just to be safe, toss out all the shiny baby bottles made with polycarbonate (see the list below or go to www.thegreenguide .com for safer alternatives) and replace them with glass or opaque polyethelene bottles. If you cannot tell whether your bottles contain bisphenol A, call the manufacturer's 800 number.

Safer plastic baby bottle alternatives (nonpolycarbonate products) include Evenflo (opaque or pastel), Gerber (colors), Rubbermaid Chuggables, Medela, bottles with disposable plastic inserts (Playtex Nurser, Playtex Drop-Ins), and baby bottle nipples made of silicone or contain no phthalates or other known chemical toxicants (they are lighter in color and safer).

SIPPY CUPS. Believe it or not, many brands of sippy cups contain bisphenol A. Some of the brands that do not contain polycarbonates are Avent Magic Cup; First Years Take & Toss; Gerber Color Change; and Playtex Sipster, Big Sipster, and Quick Straw.

TEETHING RINGS AND PACIFIERS. The CPSC suggests throwing away all phthalate-containing teething rings and soft toys used by infants, "as a precaution." Replace them with phthalate-free items. A number of retailers have pledged to sell only products that babies put into their mouths that contain no phthalates. These stores include Kmart, Sears, Target, Toys "R" Us, and Wal-Mart.

TOYS. In 2003 Greenpeace distributed a toy report card, grading twenty-one major manufacturers on their use of PVC and phthalates in toys and other articles for children. Top marks went to these companies: Brio, Evenflo, Gerber, International Playthings (Prime Time and Early Start), Lego Systems, Tiny Love, Rosie Hippo Toys, North Star Toys, and Nova Natural Toys and Crafts.

Wood toys are an excellent alternative to plastics. Commonly available well-known brands include: Plan, Haba, Jake's Room, Turner Toys, and Holztiger.

Water Bottles, Food Containers, and Food Packaging

When you think about your kitchen you realize the room where you prepare food is overflowing with plastic, including soda, juice, and water bottles; shopping bags; yogurt containers; cereal box liners; cling wrap; and sandwich bags. The really good news is that food packaging tends to be made of "food-grade" plastics with no known health hazards. However, everyday use, high temperatures, and contact with fats all speed up the process of phthalate and bisphenol A disbursal in products that contain these ingredients.

TIPS FOR PLASTIC IN THE KITCHEN

- Don't use plastic containers in the microwave. Chemicals are more likely to be released into food when plastic is heated. Glass is a safer option.

- If a number 3, 6, or 7 is in the recycling triangle at the bottom of a plastic food container, it is made from plastic that should be avoided. Instead choose products in containers with 1, 2, 4, or 5.

- Don't use cling wrap, especially in the microwave. Cover food with a paper towel or wax paper instead. If you do choose to use cling wrap, don't let it touch your food. Wrap it first in wax paper.

- Remove instant meals from plastic wrappers and trays before nuking them. If your deli food is sealed in plastic wrap, cut off a thin layer of food where it touched the wrap and next time ask the clerk not to use plastic wrap.

- When a plastic container starts to look scratched or cloudy, recycle it if possible or throw it out.

- Use something other then plastic whenever possible. Bring reusable cloth bags or boxes to the supermarket when buying groceries. Replace plastic lunch boxes with stainless-steel alternatives available at Asiana West (www.asianawest.com).

- At the market, choose metal and glass containers whenever possible. At home, store leftovers in glass or stainless steel.

- Avoid buying water bottled in plastic. Believe it or not, bottled water is less regulated then tap water. If you have concerns about your water, get it tested and/or install a filter. You can also buy one of the filter pitchers on the market. Then buy your own bottle, made from stainless steel (there are many different brands available on the Internet) or plastic that does not contain harmful chemicals, for carrying the water you have filtered at home with you.

- Be careful if you are using a polycarbonate water bottle, like Nalgene or other sports bottles. If you do use a polycarbonate water bottle, do not pour warm or hot liquids into it. The warm liquid can cause harmful chemicals to leach into your drink. If your sport bottle becomes scratched or cloudy, your drink becomes more vulnerable to the leaching of phthalates or bisphenol A.

Cling wrap and sandwich bags that are phthalate-free and safe for your food include Glad cling wrap, Glad Snap Lock bags, Hefty baggies, Hefty OneZip slider bags, Saran Cling Plus wrap, and Ziploc bags.

Sofas, Chairs, Rugs, and Electronics
You would think there would be very few phthalates in the living room and office, but in fact, phthalates are used in no-skid backings on rugs and carpets and in the upholstery on couches and armchairs. While many companies—including Sony, Ikea, Nike, Herman Miller, Shaw Carpets, Interface, Samsung, and Panasonic—are committed to phasing out phthalates, much work remains to be done to completely eradicate these toxins from our households.

Shower Curtains
Plastic shower curtains are often cited as DEHP culprits. A good option for replacing the soft plastic that keeps water inside our tubs and shower stalls is a nylon or PEVA curtain, available at Target and Ikea.

Cosmetics and Personal-Care Products
Makeup and hair and body products are also a source of phthalates. The good news is that, thanks to the European

Union's 2005 ban on DEHP and DBP in personal-care products, international cosmetic companies like Revlon and L'Oréal are now making phthalate-free cosmetics. In addition, a number of domestic beauty product companies have responded to the Campaign for Safe Cosmetics—an aggressive effort launched by a consortium of environmental groups to eliminate toxic chemicals from cosmetics, lotions, and hair care products. Some of the better-known companies that have signed on to the campaign include Revlon, L'Oréal, Kiss My Face, the Body Shop, Urban Decay, Burt's Bees, Aveda, and Aubrey Organics.

To ensure that your teenagers—whose hormonal systems are undergoing a growth spurt—are not using makeup or personal-care products that contain phthalates, visit the

Q&A:
Dr. Frederick Vom Saal

Q: Should parents worry about phthalates in their teenage daughter's beauty products?

Dr. Vom Saal: Yes. The Environmental Working Group (www.ewg.org) and Environmental Health News (www.environmentalhealthnews.org) websites offer excellent information on this.

Environmental Working Group's website and download a list of safe beauty products.

FOR FURTHER INFORMATION

Phthalate-free beauty products: www.ewg.org and www.safecosmetics.org

Phthalate-free toys and products for the home: www.cleanproduction.org and www.greenpeace.org

Information on PVC in medical devices: www.noharm.org

Quick reference on plastic products: www.thegreenguide.com and www.realmama.org

5

PCBs AND FLAME RETARDANTS

OLD AND NEW CARCINOGENS

The fourth threat we should all know about is posed by two related families of chemicals: polychlorinated biphenyls (PCBs), a class of banned compounds that linger in the environment and in our bodies for decades, and brominated flame retardants, which are used in literally thousands of consumer products, including our children's pajamas.

THE BAD NEWS

- Although the EPA banned the production of polychlorinated biphenyls (PCBs) in 1977, without a program to effectively remove PCBs from older electrical equipment, landfills, and waterways, these pollutants will persistently creep into our food chain.

- PCBs are not only persistent but they are fat soluble, meaning they can linger for decades in the fatty tissues of virtually every living organism on the planet.

- Because they accumulate in our bodies more quickly than they are excreted, the older we are when we have children, the more PCBs we expose to our babies, both in utero and through our breast milk.

- PCB levels in breast milk decrease as we breast-feed our babies—as our own decades of exposure are taken up by our infants when they consume our breast milk.

- Although PCBs are steadily declining in our fatty tissues, they are rapidly being replaced by what some news reports have called "the next PCB"—chemical compounds called PBDEs, which are found in fire retardants.

- Structurally and toxicologically like their long-ago-banned chemical cousin, PBDEs are also persistent and bioaccumulative.

- Although there have not been a great deal of studies on the effects of PBDEs, scientists believe they harm children, as PCBs do, impairing learning, memory, attention, and behavior.

- Flame retardants in fabrics such as children's pajamas are volatile, steadily releasing into the air.

- PBDE concentrations in North American women's breast milk has been measured at seventy-five times higher than levels in the breast milk of European women.

- In animal studies, PCBs and PBDEs are linked to:
  - Low birth weight
  - Decreased intelligence
  - Problems with short-term memory
  - Attention deficit disorders
  - Impaired immune function
  - Hypothyroidism
  - Various cancers
  - Disruption of sex hormones

THE GOOD NEWS

- You can greatly reduce the amount of PCBs and brominated flame retardants your children consume by serving low-fat dairy products and removing fat from meat.

- We know that farmed salmon and fish that are high up on the food chain contain industrial contaminants in their fatty tissues. Simply by removing these fish from your children's diet you will reduce their exposure to PCBs and PBDEs.

- Many corporations now provide PBDE-free products, including Ikea, Volvo, and IBM. (For a comprehensive list of companies that manufacture goods free of PBDEs go to www.thegreenguide.com.)

THE SCIENCE

Study: Scientists documented PCB toxicity in humans following accidental exposures

Scientists discovered that in 1968 pregnant women in Japan (and in Taiwan in 1978) had unknowingly exposed their babies in utero to PCBs by consuming cooking oil accidentally laced with this mixture of poisonous chemicals. Not only were there an above-average number of prenatal deaths, but the children who were born developed deficits in a wide array of learning skills crucial to success in school, including below-normal intelligence, developmental delays, and behavioral problems. Scientists followed these children through the years and learned that there was a persistent delay in growth, including reduced penile length in boys at age eleven to fourteen, and weakened immune systems. The effects were suffered not only by the children exposed in utero at the time these mothers actually used the cooking oil, but by children born to these mothers years after exposure.

Study: Effects of PCBs on Lake Michigan children

In the early 1980s some women living close to Lake Michigan consumed on average two to three local PCB-tainted fish meals each month while they were pregnant. At the time, they did not know the fish were tainted with PCBs, nor were they equipped with information about the prenatal effects of PCBs on unborn babies.

In an effort to understand prenatal exposure to PCBs, the husband and wife team of Drs. Sandra and Joseph Jacobson, psychologists at Wayne State University in Detroit, tested 212 of the children born to Lake Michigan mothers in 1980 and 1981. To identify lasting effects, they tested these same children eleven years later.

They found that children whose mothers' PCB levels (measured in blood samples and breast milk) were slightly higher than the general population had elevated rates of "low normal" IQ scores, difficulty with concentration, poor reading comprehension, and short-term memory problems. Exposure to these chemicals after birth in breast milk did not appear to further harm the children's mental capabilities. Some researchers believe this is because breast milk has built-in qualities that protect babies from environmental contaminants.

"How do PCBs and PBDEs harm our children?"

David Carpenter is the director of the Institute for Health and the Environment. A lean man with a quick mind, Carpenter's corner office is cluttered with memorabilia from numerous travels throughout eastern Europe and Asia, where he has spoken about the impact of environmental pollutants on children's health. Much of his work has focused on PCBs. He explains that like lead and mercury, PCBs are particularly toxic to the fetal brain, affecting both short-term memory and a child's ability to concentrate for long periods of time. Scientists like Dr. Carpenter have figured out that PCBs harm the fetal brain by affecting a mother's thyroid function, and they suspect that PBDEs may do the same.

Located in the throat, a mother's thyroid gland secretes thyroxine (T4), a hormone that is crucial to fetal brain development. Until babies in the womb develop their own gland, at about midgestation, they rely on their mother's thyroid hormones, which, attached to carrier proteins, are delivered to the fetus. Scientists believe this powerful ability to interfere with thyroid function at a crucial time of brain development is the means by which prenatal PCB exposure smothers human intelligence, dousing a child's ability to effectively learn and pay attention.

Additionally, the majority of scientific studies report that exposure to PCBs cause cancer. The World Health Organization, the National Cancer Institute, and the Environmental Protection Agency have all categorized PCBs as a probable carcinogen.

"PCBs were banned years ago. How is it that children are still exposed?"

Before the government banned PCBs, they were widely used as lubricants and coolants in electrical equipment such as

Q&A: Dr. David Carpenter

Q: Should women avoid eating fish while pregnant?

Dr. Carpenter: If you wait until you are pregnant to alter your diet it is too late. The half-life of methyl mercury in the body is about seventy days. [See chapter 3, "Mercury."] This means that if you want to get pregnant and to avoid the possibility that the methyl mercury will harm the intellectual development of your child, you should stop eating fish with high mercury concentration at least one year before getting pregnant. For POPs [persistent organic pollutants—shorthand for PCBs and PBDEs] the situation is even worse. The half-life of most POPs is on the order of ten years. Therefore even little girls who eat POP-contaminated fish will still have much of it in their bodies when they get to be of reproductive age. My advice: if you are female, avoid POP-contaminated fish from birth to menopause.

transformers and capacitors. Persisting for decades, PCBs enter the environment from waste sites, illegal dumping, and leaks from old PCB-containing electrical equipment. Just like mercury, when PCBs are heated up they become vapor, wafting around the planet and finding their way to chilly regions of the earth. Once they are cooled down, PCBs rain into rivers, lakes, and oceans, where they enter our food chain. With each link up the ladder—as big fish devour little fish—PCB concentrations become more potent.

Ninety percent of a child's PCB intake is through food, with fish accounting for the highest exposure. Because farmers use fish meal, fish oil, and waste animal fats as food supplements, the beef, pork, chicken, and dairy products we buy may also include a dose of PCBs. PCBs are so persistent and accumulative that those consumed by our children may actually be passed on to our grandchildren.

Q&A: Dr. David Carpenter

Q: Does grass-fed beef have a lower concentration of PCBs and PBDEs?

Dr. Carpenter: Since most POP exposure to farm animals comes from animal fats contained in the food they are fed, grass-fed beef is likely to have a much lower concentration of POPs. Not only that, but grass-fed beef has a much higher concentration of healthy omega-3 fatty acids, which are documented to reduce the risk of death following a heart attack.

Study: PCBs in farmed fish

In 2001 and 2002, Dr. Carpenter participated in a study where scientists analyzed about 460 whole farmed salmon from locations all over the globe. In addition, they purchased fish from markets in European and North American cities and obtained five species of wild Pacific salmon. Fish were weighed and filleted, and then tested for contaminants, using the US EPA methods for analyzing fish.

The study determined that levels of PCBs are much higher in farmed salmon than they are in wild Pacific salmon. Wild salmon have fewer PCBs because they eat lower on the food chain, meaning they consume fewer PCBs from other fish. They are also more active than farmed fish and therefore have less fat that stores PCBs. The report concludes, "The fact remains that salmon, especially farmed salmon, contain higher levels of these contaminants than almost any other food."

"Is my baby getting PCBs in my breast milk?"

Odds are, the answer is yes. In the late 1960s a Swedish researcher found PCBs in the tissue of a dead eagle. Soon after, on a hunch that these compounds were taking up residence in human fatty tissues, he analyzed his wife's breast milk and was startled to learn that even the most sacred food was contaminated with this toxic mixture of chemicals. Soon after, Sweden banned PCBs and in 1977 the United States did the same. However, because they tenaciously hang out in our bodies for decades, even women

Q&A:
Dr. David Carpenter

Q: Is there any way to prepare fish so that it contains a lower concentration of PCBs?

Dr. Carpenter: There are some things that can be done, such as grilling so that the fat drips off and is not consumed. Also removing the skin and the dark streak, which contains a lot of fat, helps some. However, our recent study has shown that skin removal only reduces the PCBs by about 25 percent.

born long after the 1978 ban on PCBs are mainlining these persistent pollutants to their babies in breast milk. However, it is important to remember that breast milk contains powerful ingredients that in study after study have been shown to not only keep children healthy but make them smarter.

"Given that breast milk has high levels of contaminants, should we bottle-feed our babies?"

The research strongly says no. Although breast-fed babies test higher for PCBs than babies exposed only in utero, as a group they consistently perform better then formula-fed babies. Indeed, duration of nursing is positively related to health, memory, and language—in other words, the longer kids nurse, the better off they are. Researchers surmise that this is because breast milk is actually formulated to protect our babies against harmful chemicals. The World Health Organization and the American Academy of Pediatrics continue to recommend at least one year of breast-feeding. Unfortunately, we will never know how much better breast-fed babies would perform without the chemical additives contaminating their precious food source. In our PCB-laced world there is no way to find a control group.

"How exactly do flame retardants affect our children?"

PBDEs are a group of compounds used as flame retardants in children's pajamas, computers, TVs, foam for furniture, upholstery, rugs, draperies, and car interiors. In most cases, these chemicals impede ignition, and if a fire starts, they hinder the spread of flames, allowing people time to escape. Of the 209 PBDE compounds, deca-BDE, octa-BDE, and penta-BDE are the most common. Proven to be the most bioactive, penta and octa are no longer manufactured in this country. As products that contain PBDEs age and break down, these chemical mixtures seep into the air. Like PCBs, flame retardants stick around for a long time, can travel far, and accumulate in the fat and tissue of humans, animals, and fish.

It is difficult to detect the damage PBDEs do to our children's growing bodies. Scientific research conducted on rodents has revealed that, like PCBs, PBDEs interfere with the thyroid gland's ability to send its hormone thyroxine to do its job in promoting a healthy fetal brain. Because

PCBs and PBDEs are structurally and toxicologically similar, scientists anticipate seeing all of the cognitive and health problems now attributed to PCBs.

"So how about kids' pajamas?"

It is a terrible dilemma. In the United States most children's pajamas are treated with flame retardants, which can give children critical extra moments to escape a fire. Flame-retardant sleepwear is likely to be labeled as "flame resistant." An alternative to flame-resistant pajamas is sleepwear that has the label "Wear snug-fitting. Not flame resistant." The idea is that if no air is allowed to circulate between the fabric and a child's skin, it is less likely to catch fire. My kids sleep in snug-fitting, cotton long underwear that is not treated with flame retardants. Because it fits close to the skin, little air circulates. This provides some protection from fire without the toxic chemicals.

> "TOXIC BREAST MILK: A nursing mother takes
> a hard look at nature's perfect food."
> —*New York Times* (January 9, 2005)

"Can our babies get PBDEs in breast milk?"

It was the Swedes who first detected PBDE in mothers' breast milk by analyzing stored supplies from early 1972 to 1997. The discovery that levels of PBDEs were doubling in breast milk every five years led to the Swedish government's ban on some PBDEs by July 2003. Since the ban, PBDE levels in Swedish mothers' breast milk have decreased significantly.

Study: PBDEs in breast milk

Dr. Arnold Schecter from the University of Texas conducted the first-ever comprehensive study of PBDEs in mother's milk in 2002. By analyzing forty-seven samples of milk from two Texas milk banks, he determined that levels were between ten and one hundred times higher in the United States than in Europe, where PBDEs are used less. In fact, median blood levels found in the US population show an exposure of PBDEs similar to that of Swedish laborers who actually worked in factories that manufactured deca-PBDE–treated rubber. US levels are far higher than those that warranted banning the use of PBDEs in products in Sweden.

"Do PBDEs pass from mother to baby in utero?"

Similar to PCBs, PBDEs circulate in blood that is pumped through to our babies in utero. A study by Anita Mazdai and others at Indiana State University compared samples of twelve mothers' blood, taken when they were admitted for childbirth at two Indiana hospitals, with samples of their newborns' cord blood. Mazdai learned that there was a near perfect correlation between a mother's level of PBDEs and that of her baby in utero, showing that PBDEs readily cross the placental barrier.

Q&A: Dr. David Carpenter

Q: Given everything we know about PCBs and PBDEs, should we breast-feed our babies?

Dr. Carpenter: There is overwhelming evidence that breast-feeding has important benefits, improving immune function in the child, protecting against chronic diseases even when the child grows to adulthood, and promoting bonding between the mother and the child. While the presence of these contaminants in breast milk is not a good thing, under almost all circumstances breast-feeding has greater benefit than risk.

Study: PBDE levels higher in children's bodies

For one small study, a single Berkeley, California, family of four—mom, dad, a five-year-old daughter, and an eighteen-month-old son—all gave blood to be tested for levels of PBDEs, first in September 2004 and then again ninety days later. The Berkeley family used no common household cleaners or pesticides, had no wall-to-wall carpeting, and owned no new large appliances. The mother was a university researcher and the father, who was from the east coast, taught high school. Their daughter attended kindergarten and the baby, who was still breast-feeding at the time of the study, spent his days both in child care and at home with his mother and father.

The researchers in this study discovered that, unlike PCBs, which accumulate in our bodies as we age, levels of PBDEs are two to fifteen times higher for children than adults. Most alarming was that these levels were uncomfortably close to those associated with adverse effects on reproduction and neurodevelopment in laboratory animals. While the adults' PBDE concentrations approached US median concentrations, the children's concentrations were near the maximum found in American adults.

So why were the children's levels sky high when the parents' levels were average? One explanation is that because the children spent a good deal more time on the floor they had more exposure to the penta- and deca-PBDEs in the household dust. Household dust accounts for 80 percent of total daily PBDE exposure for toddlers, compared with 14 percent for adults. The EPA estimates that children from age one to four ingest one hundred milligrams of household dust per day and adults ingest fifty milligrams per day.

The children's diets could be another source of exposure. Although there is a lack of data on dietary PBDE exposure, there is good data on exposure through breast milk. Dr. Schecter assumes that although both children were breast-fed, the baby's level was particularly high due to more recent breast-feeding. "Babies get a whopping dose from mother's milk," he said. As mothers who provide this magic

elixir, mother's milk, we have to wonder at its almost cura-
tive power in offsetting the dangers of this environmental
toxin.

"STUDY REVEALS TOXIC CHEMICALS IN
HOUSEHOLD DUST"
—*Wall Street Journal* (March 23, 2005)

Study: PBDEs in household dust

The Environmental Working Group (EWG), a watchdog
environmental agency in Washington, DC, collected household dust from ten homes in states across the country. Testing these samples, they learned that the levels of brominated
fire retardants (PBDEs) found in dust bunnies under beds
and behind sofas is unusually high compared to the other
chemical compounds they tested for. Prior to the dust study,
they had tested breast milk from ten of the twenty study
participants. Comparing these studies, they were mystified
to learn that there appears to be no correlation between
PBDE levels in household dust and PBDE levels in our bodies. One possible explanation might be that some people
tend to absorb more PBDEs than others, either metabolizing
them differently or eliminating them more slowly.

WHAT YOU CAN DO TO PROTECT CHILDREN FROM
PCBs AND PBDEs

PCB levels in blood have been steadily decreasing since the federal government banned these industrial chemicals in 1977. On the other hand, PBDEs are skyrocketing. Here are some steps we can all take to protect our children from exposure to PCBs and PBDEs:

- Farmed salmon contain high levels of PCBs and PBDEs; all salmon labeled "Atlantic" are farmed. If you feel you must feed salmon to your children, select wild-caught salmon instead, or opt for fish that are lower on the food chain—for instance shrimp, tilapia, and flounder. Do keep in mind that some researchers suggest that pregnant women and breast-feeding women should avoid fish altogether. For a list of fish that are safe to eat, see page 60—PCBs and PBDEs accumulate in a fish's body the same way that mercury does.

- PCBs and PBDEs build up in the fatty tissue of meat and dairy products. Therefore it makes a lot of sense to reduce the amount of fat in the food you serve your children. For instance, choose skim milk, low-fat yogurt, skinless chicken breast, and turkey. Because they do not consume animal fat in their feed, grass-fed animals and fowl contain fewer PCBs and PBDEs. To buy grass-fed products go online to: http://eatwild.com. There you will find a wealth of information about the farmers as well as information on how and where to buy their products.

- High levels of PBDEs have been found in household dust. One way to limit PBDE exposure to small children who spend a lot of time on the floor is to vacuum floors and upholstered furniture regularly with a vacuum that has a high-efficiency, or HEPA, filter.

- Wipe dusty surfaces with a wet cloth and mop floors regularly.

- Purchase furniture, children's pajamas, electronics, cars, and carpets from companies that have chosen not to use PBDEs in their products. The Green Guide at www.thegreenguide.com lists some companies that make products free of PBDEs:

- Pajamas: Patagonia, Ecoland, Under the Nile, and Garden Kids
- Furniture and cribs: Ikea, Berkeley Mills, Lifekind, Gaiam, Ecobaby, Organic Bliss Innerspring Mattress, Natura Sleep Systems, Natural Aurora Mattress, Abundant Earth, Organic Cotton Alternatives, and Tonkatinkers Kreations
- Electronics: Sony, Motorola, and Intel (PBDE-free), HP monitors (PBDE-free), Apple, Canon, Hitachi, Panasonic, NEC, and Toshiba (reduced PBDEs)

■ Write or call manufacturers to let them know that you want them to stop making products with PBDEs and that you will not buy their products until they do.

FOR FURTHER INFORMATION

www.ewg.org/reports/inthedust/summary.php

www.watoxics.org/content/pdf/PBDEsFactSheet.pdf

www.atsdr.cdc.gov/tfacts68.html

www.tcodevelopment.com

www.svtc.org

http://e.hormone.tulane.edu

www.nrdc.org/breastmilk/default.asp

www.epa.gov/opptintr/pcb

www.thegreenguide.com

http://healthychild.org

6

AIR POLLUTION

DIRTY AIR HARMS OUR CHILDREN'S BRAINS

Air pollution may seem like too big a problem to take on: sure, air pollution is bad, but what can I possibly to do about it? However, much of the worst air pollution is very local in origin, like fumes from diesel engines in school buses. Given that childhood exposure to air pollution is linked to a host of chronic, lifelong respiratory ailments, you should definitely be aware of the critical steps you can take to protect your child.

THE BAD NEWS

- Researchers tracking 3,500 students in southern California determined that there was an increased onset of asthma in children involved in outdoor activities in communities with high levels of ozone, or smog.

- Air pollution reduces children's head circumference and lowers IQ or, as a memorable *New York Post* headline put it, reporting the result of one study, "Air More Stinky, Kids Less Thinky."

- There is a slight increase of sudden infant death syndrome (SIDS) in cities with high levels of fine particulate matter (PM) pollution.

- Air pollution has also been linked to:
  - Low birth weight
  - Premature births
  - Retarded lung growth
  - Cancer

THE GOOD NEWS

- Air pollution is trending down.

- Instruments for detecting the harmful effects of air pollution have gotten much better, offering insight into how we can best protect our children from the soup of chemicals in the air we breathe.

- Many key sources of air pollution are very local, from tobacco smoke in the home to school buses idling at schools. There is a lot you can do to eliminate or at least reduce them.

- Some days are far worse than others—you can reduce your child's exposure by watching the weather and keeping kids inside during the hottest hours on high-ozone days.

THE SCIENCE

There are a few obvious reasons why children are vulnerable to air pollution. To begin with, their small airways are very susceptible to closing up when irritated or inflamed. They also breathe more rapidly than adults, inhaling higher doses of air pollutants per pound of body weight. Finally, because children have seemingly boundless energy, they often spend a lot of time running around and playing outdoors, particularly on hot, sunny days when smog is at its worst.

To understand air pollution, one first has to become familiar with a few of the most harmful and common culprits affecting our air: particulate matter (PM), polycyclic aromatic hydrocarbons (PAH), and the ozone, otherwise known as smog. The American Lung Association has found these hazardous polluters to be widespread all across the country, and I'll be referring to them throughout this chapter.

PM (Particulate Matter)

Particulate matter—also called particle pollution—is a catchall term for the most dangerous specks of airborne pollution that we breathe. It is usually created by burning. Major sources of particulate matter are diesel trucks and buses; cars, especially SUVs; inefficient home heating systems; and inefficient home fireplaces—both indoor and outdoor. Sometimes composed of microscopic particles, PM can go so deep into the lungs that no amount of coughing will dislodge it. Some PM is so fine that it exists as vapor and effectively passes through the lungs into the bloodstream.

PAH (Polycyclic Aromatic Hydrocarbons)

PAHs are a family of more than one hundred chemicals formed during the burning of gas, diesel fuel, oil, coal, wood, tobacco, and garbage—that is to say, from all motorized vehicles, power plants, cigarette smoke, and even our backyard barbecues. Our children are exposed to PAHs when they breathe smoke from any of these sources. In the body, PAHs are transformed into chemicals that attach themselves to chromosomes. These attachments, called adducts, can be counted in white blood cells, allowing researchers to correlate exposure to PAHs with cancer.

Ozone

One minute you hear that we are destroying the ozone layer and the next you're advised to stay inside because it is a "high-ozone day." The confusion lies in the fact that there are two different kinds of ozone. One kind, up in the stratosphere, protects us from the sun's ultraviolet rays. This is the *ozone layer*. Depletion of the ozone layer affects global warming.

The other kind of ozone is commonly referred to as smog—the toxic haze you see over some cities. This ozone is formed by a chemical reaction when raw ingredients from tailpipes, smokestacks, gas stations, paint, refineries, and chemical plants come into contact with heat and sunlight. For instance, when you fill your car up with gas on a hot, sunny day, the vapors from pumping the gas combine with the heat and sun to contribute to the layer of smog.

"CANCER IS AIR 'BORN'"

—*New York Post* (February 16, 2005)

"How does air pollution affect our children?"

Dr. Frederica Perera, a molecular epidemiologist who launched the Columbia Center for Children's Environmental Health, measures levels of prenatal exposure to air pollution, then records how exposure specifically harms our babies in the womb. In various studies conducted in Poland and New York City, she has focused on PAHs because these compounds attach themselves to chromosomes and can be counted in white blood cells. By counting the number of attachments, called adducts, Perera is able to correlate

PAH exposure to abnormal chromosomes—which cause cancer—a direct link between air pollution and injury.

Dr. Perera first began testing the white blood cells in umbilical cord blood because she was looking for a clean control group—surely newborns had not yet had time to breathe in dirty air. However, she was alarmed to discover that even children whose mothers were nonsmokers had detectable levels of these chromosomal abnormalities. Her research also showed that a baby has more chromosomal damage per estimated level of exposure than its mother, illustrating that the fetus lacks the ability to ward off toxins in air pollutants. Dr. Perera explained, "The higher the hydrocarbons in a mother's air, the more frequent the abnormalities seen in an infant's chromosomes."

Study: Air pollution makes our babies smaller at birth
Committed to unraveling the long-term health effects of air pollution in young children, including growth, cognitive development, and cancer, Dr. Perera launched the massive,

"PAHs found in the particulate form can act like time-release capsules in the lung. Once they travel to the deep lung, they are rapidly absorbed into the bloodstream, where they are transported systematically."—Dr. Frederica Perera

long-term Mothers and Children's Study in 1998. Seven hundred expectant African-American and Dominican women living in northern Manhattan and the South Bronx were recruited for the study. To record the levels of prenatal exposure to air pollution, including particulate matter, diesel exhaust, polycyclic aromatic hydrocarbons (PAHs), pesticides, secondhand tobacco smoke, and home allergens (mouse, cockroach, and dust mite droppings), these soon-to-be mothers, after careful interviews, wore backpacks that contained compact air monitors. Once a participant felt her first labor pain she was to call a representative from the Center for Children and Environmental Health, who would meet her at the hospital to retrieve the placenta and draw blood from both the mom and the umbilical cord.

From 2000 to 2003, 263 expectant mothers in the New York City cohort participated in an investigation that correlated exposure to air pollution with fetal injury.

Q&A: Dr. Frederica Perera

Q: What are the primary sources of ambient PAHs?

Dr. Perera: Fossil fuel and wood combustion are the primary sources of ambient PAHs.

Dr. Perera's findings showed that full-term babies whose mothers were exposed to the highest levels of air pollution were significantly smaller in weight, length, and head circumference—even after accounting for height and weight of the parents, secondhand tobacco smoke, gestational age, sex of the baby, and season of birth. These findings are problematic because reduced birth weight, in particular, has been linked to delays in cognitive development as well as increased risk of type II diabetes, hypertension, and coronary heart disease during adulthood.

Study: Increased SIDS in cities with high fine particle pollution

When studies began to suggest that sudden infant death syndrome (SIDS) could be linked to secondhand cigarette smoke, a research team led by Tracey J. Woodruff wondered whether there was also a potential association between SIDS and particulate air pollution. In 1997 these researchers took

There was an increase in the number of small babies born around the World Trade Center soon after the terrorist attack. Researchers from Mount Sinai School of Medicine surmise that it was probably a result of air pollution from the smoldering disaster site.

a close look at information on SIDS-related deaths, from the National Center for Health Statistics (NCHS), and compared it to EPA data about PM air pollution. After taking into account factors that might affect their findings, such as a mother's smoking during pregnancy and socioeconomic differences, these researchers learned that babies who were exposed to high levels of PM were at a slightly higher risk of SIDS than babies who had very little exposure.

"How do I know if my town has a lot of air pollution?"

To figure out which US residents are at the greatest risk for health problems from poor air quality, the American Lung Association assessed degrees of air pollution in terms of both particulate matter and ozone. As you can see, several cities turned up on both lists, giving their residents a double whammy of air pollution.

In terms of year-round *particle* (PM) pollution, these cities and urban corridors ranked the worst (listed in descending order of dirtiness):

Los Angeles–Long Beach–Riverside, CA
Bakersfield, CA
Pittsburgh–New Castle, PA
Visalia–Porterville, CA
Fresno–Maderas, CA
Detroit–Warren–Flint, MI
Hanford–Corcoran, CA
Cleveland–Akron–Elyria, OH

Birmingham–Hoover–Cullman, AL

Atlanta–Sandy Springs–Gainesville, GA

In terms of *ozone* pollution, these cities ranked the worst (in descending order):

Bakersfield, CA

Los Angeles–Long Beach–Riverside, CA

Visalia–Porterville, CA

Fresno–Madera, CA

Merced, CA

Houston–Baytown–Huntsville, TX

Sacramento–Arden–Arcade–Truckee, CA

Dallas–Fort Worth, TX

New York, NY–Newark, NJ–Bridgeport, CT

Philadelphia, PA–Camden, NJ–Vineland, NJ

Asthma: The Most Pervasive Chronic Disease among Children

You cannot talk about air pollution without considering the rapid increase in asthma in developing countries throughout the world. More and more, parents with no

The CDC estimates that 6 million American children have asthma.

From 1980 to 2003, the prevalence of asthma in American children rose from 3.6 percent to 5.8 percent, about 60 percent higher.

history of asthma are shocked when their children develop the tenacious, chronic disease. What causes asthma? Cases of asthma have risen too quickly to be explained away by slow-moving genetic changes in the human population. Therefore, like so many chronic diseases, researchers suspect asthma is the result of a genetic susceptibility coupled with environmental contaminants.

"Have researchers linked specific environmental factors to asthma?"

Scientists are getting closer to understanding the role that toxins in the air play in asthma. Here is some of the evidence they have compiled:

- One study conducted by the University of Southern California followed 3,535 children for five years in southern California neighborhoods. Researchers looked at new cases of asthma to determine whether or not onset could be connected to growing up in communities with dirty air. Six of the neighborhoods had polluted air, while the other six had measurably better air quality. Analyzing these children, researchers determined that ozone strongly contributed to asthmatic conditions in children who previously did not have the disease. Nationwide, the highest increases in asthma are in urban areas that have higher ozone levels.

- In communities with high ozone concentrations, children who play a lot of outdoor sports, breathing these chemicals deeply into their lungs, have a 330 percent higher risk of developing asthma than children who do not play outdoor sports. In areas with low ozone concentration, playing sports outdoors has no effect. In general, time spent outside in high-ozone areas contributes to a 40 percent higher incidence of asthma.

- Some researchers report that asthma is becoming more pervasive as a result of depletion of the ozone layer. As the climate gets warmer, there is an increase in smog and other impurities, including allergens like mold and pollen that interact with air pollution, giving rise to even more asthma.

- Secondhand cigarette smoke has been consistently linked to an increased frequency and severity of new-onset asthma attacks in children. If parents smoke in the home, their children's risk of developing asthma at least doubles. This also puts the children at risk for other respiratory problems, including bronchitis and pneumonia, and, later in life, for lung cancer.

- Tobacco exposure in the womb is also very harmful. If a mother smokes while she is pregnant her baby's asthma risk triples. A host of additional dangers for children exposed to secondhand cigarette smoke in the womb include premature birth, low birth weight, and SIDS.

- Breast-fed babies are less likely to develop asthma and allergies than babies who are fed formula. Ample scientific evidence shows that breast milk boosts a baby's immune system, decreasing susceptibility to allergies.

- Researchers have discovered that only some babies in the womb become sensitive to environmental contaminants that cross the placenta. Those babies emerge into the world with a strong predisposition to both allergies and asthma.

INDOOR ALLERGENS LINKED TO ASTHMA

Pet dander

Dust mites

Cockroach feces

Fungal contamination (mold)

Secondhand cigarette smoke

When you read this list you'll see that most of these allergens can be controlled or limited by you. Not as easy to control is how air pollution affects children whose breathing is already compromised.

OUTDOOR ALLERGENS LINKED TO ASTHMA
   Pollen
   Ozone
   Nitrogen oxide
   Particulate matter (PM)
   Diesel exhaust
   Some pesticides
   Some types of wood dust

If your child suffers from asthma, you'll want to have his particular allergens identified by a specialist. Perhaps it will be possible for your child to avoid many of these outdoor allergens without other medical treatment. In any case you'll want to follow your doctor's recommendation for avoiding and dealing with emergency situations.

> "AIR MORE STINKY, KIDS LESS THINKY"
> —*New York Post* (May 30, 2006)

Study: Air pollution affects how well our children learn

In the early 1990s, the seven hundred children who lived in New York City and participated in Dr. Perera's longitudinal study were three years old. Curious to know if air pollution doused human intellect, Perera and her researchers used a standardized test of mental and psychomotor development to evaluate these children. They then compared each child's facility to learn with his or her level of PAH exposure in the womb. Of course, investigators controlled for other exposures that might have skewed their findings, such as socioeconomic factors and exposure to tobacco smoke, lead, and other environmental contaminants. What Perera and her team found is that exposure to PAHs in utero could not be definitively linked to motor development or to behavioral problems. It was, however, linked to lower scores on mental development. And unfortunately PAHs are not easy to avoid. As Dr. Perera explains, the kind of urban air pollutants examined in the study "are very pervasive throughout much of the city."

"We live on a busy road. Could pollution from trucks and cars be contributing to my son's bronchitis?"
It seems reasonable to conclude that kids who live and attend school closer to streets where trucks, buses, SUVs, and cars are frequently racing by or jammed in traffic are breathing in more PM, PAHs, and ozone. A number of studies confirm this reasoning:

- In a 2005 study Dr. Rob McConnell of the Keck School of Medicine, at the University of Southern California, found that the children who lived within 75 meters (about 82 yards) of a heavily trafficked street or freeway were nearly 50 percent more likely to have had an asthma episode in the past year than kids who lived more than 300 meters (about 328 yards) away from a major roadway.

- A study of thirty-nine thousand children living in a city in Italy found that the kids who lived near busy roads were 60 to 90 percent more likely to become sick with various respiratory illnesses, including bronchitis and pneumonia.

- A study of more than 3,700 German preteens discovered that those who lived near streets frequently used by trucks were 71 percent more susceptible to hay fever and 50 percent more prone to wheezing.

Study: Children are exposed to diesel exhaust on school buses

Because of diesel exhaust, the air inside a school bus can be substantially more polluted than the air outside the bus. The exhaust from diesel school bus engines is made of very fine particles of carbon as well as other toxic gases. There is no known safe level of exposure to diesel exhaust for children, especially those with respiratory illness.

In 2003 researchers from the Berkeley and Los Angeles campuses of the University of California boarded seven school buses and tested the inside ambient air while they were driven around the vast city of Los Angeles. All the buses they tested, except one 1975 model, were manufactured between 1985 and 2002. They learned that children riding to and from school in these buses inhaled 34 to 79 percent more air pollution than the average weekday commuter did on the same day. The worst in-bus air was in the older buses driving in heavy traffic with the windows closed.

VARIABLES THAT CONTRIBUTE TO DIRTY AIR INSIDE SCHOOL BUSES

Buses idling while children gather to board or de-board

Ventilation inside buses or whether windows are open or closed

Bus routes

Engine model

Age of the school bus

Bus maintenance cycles

"I have heard that idling the engine actually produces more dirty air inside and outside a school bus than driving the bus does. If this is true what can parents do to ensure better air quality for children riding buses?"

Yes. Lined-up, idling school buses spew high levels of particulates. Many community leaders and parents have worked hand in hand with their schools and school boards to create policies that ensure the best quality air possible on school buses.

Real Stories

Peter Iwanowicz is the father of two children who attend public school near Albany, New York. Last year a parent at his school with an asthmatic child launched a campaign to address the health-threatening problem of buses, minivans, SUVs, cars, and trucks idling outside their children's elementary school. Working together, these parents convened a team of concerned mothers and fathers and created a list of facts that would be meaningful to administrators, including: the American Academy of Pediatrics analysis of health issues from idling school buses; fuel costs; and wear and tear on bus engines. They then presented their list of concerns to the school superintendant, the school's parent group, the school board, and the bus drivers themselves. Iwanowicz explains, "After we got everyone on board we were able to very quickly change the bus idling policy at my kids' school."

Study: Air pollution slows children's lung growth

As scientists James Gauderman of the University of California learned more about air pollution and children's health he began to wonder whether air pollution actually stunts lung development, resulting in lifelong health problems. To investigate this concern, he and a team of researchers followed 1,759 children in twelve southern California neighborhoods, examining lung capacity for eight years. When they began their study, the children were ten years old—a time when lungs embark on their final spurt of growth. At the close of the study, when the children were eighteen years old, the researchers discovered that children exposed to polluted air had lung capacities less than 80 percent of what is considered normal. The teenagers who participated in the study had never smoked and never had asthma.

What does it mean for children to have reduced lung capacity? It can translate into wheezing during a cold or a bout with the flu. Lame lungs can also make it more difficult to recover after an illness. Although southern California is notorious for its bad air, because air pollution has dropped there in the last twenty years, and because air quality has deteriorated in other cities across the country, its level of pollution is now comparable to that in most urban areas. "The solution is really at the level of government regulations," Professor Gauderman says. "You would have to put additional emissions controls on motor vehicles to control this problem."

WHAT YOU CAN DO TO
PROTECT CHILDREN FROM
AIR
POLLUTION

- Cut down on driving your car. Encourage a family practice of using bikes, feet, and public transportation. Make the extra effort to carpool and teach your children to think in terms of ride sharing. Plan ahead and combine errands into one trip.

- Do not idle your car. If you are waiting, turn off your motor. In the winter, dress warmly when you set out so that you're prepared to wait in the car without the heater on.

- Keep your car in good running order. Make sure your tires are properly inflated to cut down on fuel consumption. Have the oil and filters changed at prescribed intervals.

- Drive a low-emission car. If you're not already driving a low-emission car, and can't afford to buy one, do the research before you buy your next car and "invest" in improving the quality of our air.

- Appeal to administrators and the school board about diesel fumes. Ask them what obstacles need to be cleared to prohibit the idling of school buses, as well as other vehicles, outside your child's school.

- Designate the newest and cleanest school buses at your child's school to the longest bus routes and for all field trips. The longer children are on the bus, the greater their exposure. Though it is an imperfect solution to reduce exposure, it will minimize exposure overall, so take this up with your school administrator.

- Do not allow buses to closely follow one another. Children on school buses that travel directly behind one another will be exposed to higher levels of PM. Bus drivers should be made aware of the significant effects of following other vehicles and be required to allow a distance of at least one city block between buses. Exhaust emissions from other large vehicles, besides school buses, can contribute to poor air quality inside the school bus. Ask your school administrator to stagger departure times of buses as they leave at the end of the school day.

- Keep all diesel buses well maintained. Cost-saving programs often result in delayed maintenance—the bus still runs, but not efficiently, so it pollutes more. Ask to see the maintenance schedule for your school's buses.

- Replace your lightbulbs with energy-saving bulbs—and get your child's school to do the same. Energy saving fluorescent lightbulbs reduce your electric bill and, in the big picture, put less drain on power usage; and the less we rely on our power plants, the less air pollution we'll have.

- Look for warning signs of undiagnosed asthma. Kids with asthma are especially sensitive to air pollution. Pay particular attention to prolonged and regular bouts of coughing. Notice whether they seem to have unusual shortness of breath when playing sports or exercising.

- Make your indoor environment "lung healthy." Do not allow anyone, including yourself, to smoke cigarettes indoors. Clean regularly to reduce dust and insect droppings, fix leaks and moisture issues that might cause mold, reduce usage of woodstoves and fireplaces. Opt for cleaner stoves such as those that burn pellets and corn.

- Stay tuned to local news reports about air quality. If it is going to be a "high-ozone" day, limit the amount of time your children spend outdoors. Arrange schedules so that your children will play activities that make them breathe hard before ozone levels start to climb.

■ Keep all outdoor activities as far as possible from heavily trafficked roads. Coaches often send kids to run for conditioning along busy streets. If this is true at your child's school, you may wish to speak up about it. Make good choices about parks and playgrounds where your children play.

■ Talk to community leaders—get involved. Ask them to address particular issues that concern you. Be willing to help bring about the changes you seek.

FOR FURTHER INFORMATION

American Lung Association: www.lungusa.org

Clean School Bus Campaign (Union of Concerned Scientists): http://cleanschoolbus.org

Diesel Exhaust (EPA): www.epa.gov/ne/eco/diesel

Idling Reduction (EPA): www.epa.gov/smartway.idling.htm

Diesel Exhaust and Children (Environment and Human Health, Inc.): www.ehhi.org.diesel

Car Talk website, where the Tappett Brothers provide suggestions for environmentally friendly driving practices: www.cartalk.com/content/eco/tips.html

7

PESTICIDES

WHICH FOODS ARE SAFE?

Pesticides remain a cause for concern. In 2000 the EPA banned some of the really bad chemicals for use in homes, significantly decreasing their levels in children's bodies. However, pesticides still abound in our food and nobody knows for sure how they affect our children. Designed to kill insects, it may well be that they contain the seeds of problems that have yet to manifest themselves in a serious way. For this reason it is prudent to take steps—none of which are terribly drastic—to reduce children's exposure to pesticides.

THE BAD NEWS

- Nonorganic farmers use around two hundred approved chemicals on their crops.

- Children who eat a conventional diet are exposed every day to tiny amounts of more than thirty chemical pesticides designed to poison insects.

- The EPA estimates that about 80 percent of our total pesticide exposure comes through the food we eat. The remaining 20 percent of exposure to pesticides comes through drinking water and by using pesticides at home to control insects and rodents.

- Pesticides have been linked to
  - Cancer
  - Birth defects
  - Kidney damage
  - Liver damage
  - Reproductive disorders

THE GOOD NEWS

- In 2000 the EPA banned some organophosphate pesticides for use in homes, driving down levels of pesticides in children's bodies.

- A study has confirmed that by feeding your children organic food, you can eliminate nearly 80 percent of their pesticide exposure.

THE SCIENCE

"ORGANICS: ARE THEY WORTH IT?"
—*Eating Well* (February/March 2006)

Study: Pesticide levels reduced in the bodies of children who eat only organic food

In 2003 a group of researchers from the University of Washington measured pesticides in the urine of twenty-five Washington State children before and after they switched to organic food. After only five days on an organic diet, pesticide residues in urine decreased to undetected levels and remained so until they began to eat conventionally grown food once again. The study concluded, "An organic diet provides a dramatic and immediate protective effect against pesticide exposure." However, the researchers noted that evidence is still lacking about whether or not low-level pesticide residue in food harms children.

"Can pesticide residue on food harm my child?"
No one knows for sure whether the kind of low-level exposure our children get by eating nonorganic food is harmful.

While there is evidence from animal studies that fetal exposure to some pesticides at high doses can cause neurodevelopmental problems, the levels of pesticide exposure our children get from conventional food is much lower than that. One study, however, done by Dr. Elizabeth Guillette, an anthropologist from the University of Florida, offers some insight into how these chemicals might be affecting our children.

Study: Effects of pesticides in Yaqui children

In 1993, Dr. Guillette traveled to northwestern Mexico, where a tribe of Yaqui Indians had split up in the late 1940s, at the outset of the Green Revolution. One faction chose to stay in the valley, where fields treated with pesticides went right up to the town line. The other moved to the foothills, where they practiced traditional ranching methods free of chemicals.

Their children provided a perfect study subject in that the two groups shared a genetic background, diet, and cultural practices. After performing a series of tests on the four- and five-year-olds in both groups, Dr. Guillette learned that although there was no difference in their growth patterns, the group of children exposed to pesticides repeatedly scored lower in the tests designed to measure stamina, gross and fine motor skills, eye-hand coordination, thirty-minute memory, and the ability to draw a person.

Dr. Guillette returned to the Yaqui Valley when these same children were six and seven years old and repeated the same tests. What she saw was that the pesticide-exposed kindergarteners and first graders on the whole had an inferior sense of balance, difficulty solving easy puzzles, and poor eye-hand coordination. Perhaps related to these low ability levels, they were easily frustrated and had trouble completing tasks.

In the third phase of the study, Dr. Guillette evaluated girls in prepuberty and discovered early breast development in the children who had been raised in the pesticide-using agricultural valley. In the mostly pesticide-free foothills, the girls had some breast development by age ten with a normal correlation of fat deposits and mammary tissue. In the valley, however, there appeared to be only fat deposits, with no mammary tissue in the developing breasts of these young girls.

Dr. Guillette inferred that the girls' abnormal breast development was most likely caused in utero. Mammary gland tissue is first laid down at six to eight weeks and then again at twenty weeks. If pesticides somehow inhibited the proper development of mammary tissue, when these girls grow up and become mothers they will be unable to breast-feed their babies.

"How does the EPA assess pesticides on food and children's health?"

Congress passed the Food Quality Protection Act (FQPA) in 1996, giving the EPA ten years to evaluate all pesticides used on food. The EPA completed this task in January 2007. Their evaluations considered cumulative sources of exposure from both food and water. Additional safety factors were established in testing for the heightened sensitivity of children.

"The most meaningful changes," says Jim Jones who heads up pesticides investigations for the EPA, "were changes in organophosphate pesticides, which have been shown to be neurotoxic in animal studies. This class of pesticides is no longer allowed to be used in homes." Before 2000 these dangerous chemicals dominated the home products market.

"I have heard that fruits and vegetables from other countries are grown with chemicals that have been banned in the United States. If this is true, should I stop buying imported fruits and vegetables?"

Food imported from other countries is required by the US Department of Agriculture (USDA) to meet the same federal standards as domestic produce. The Food and Drug Administration (FDA) food surveillance monitoring at the border routinely checks for pesticides and other contaminants. There is a debate among advocates, however, as to whether or not the FDA is aggressive enough. Therefore it is probably wise to avoid fruits and vegetables grown in other countries.

"How can I feed my children organic food without breaking the bank?"

Parents tend to feed their children organic food for one reason: they don't want their kids to ingest pesticides. Although it is wise to feed your children organic food whenever you can, it is not always possible to find organic produce and it is also more expensive. The USDA regularly tests samples of organic and nonorganic fresh and processed food from markets around the United States for pesticide residues. Among other things, what they've found is that fruits contain more pesticides than vegetables. You can find their results at www.ams.usda.gov. The Environmental Working Group used the USDA's results of more than one hundred thousand tests for pesticides on produce between 1992 and 2001 to develop a ranking of contamination by type of produce (see www.foodnews.org for more on their findings). Here's what they learned:

- Highest pesticide contamination: apples, bell peppers, celery, cherries, imported grapes, nectarines, peaches, pears, potatoes, red raspberries, spinach, and strawberries. Buy organic for these.

- Lowest pesticide contamination: asparagus, avocados, bananas, broccoli, cabbage, kiwi, mango, onions,

In 2005, *Consumer Reports* magazine stated that on average consumers pay 50 percent more for organic fruits and vegetables and sometimes 100 percent more for organic meat and dairy.

papaya, pineapple, sweet corn (frozen), and sweet peas (frozen). It is not always necessary to buy organic versions of these foods.

On baby food: Although Gerber has banned most organophosphate insecticides as well as other pesticides from nearly all its products, *it is important to buy organic baby food.* Babies develop so quickly that they are particularly vulnerable to toxins. It stands to reason that since baby food is composed of pureed fruits or vegetables, it would concentrate any pesticide residues present on them.

According to the EWG, eating the twelve most-contaminated fruits and vegetables exposes you to an average of twenty pesticides a day. If you choose from the list of the twelve least-contaminated fruits and vegetables, you will expose yourself to about two pesticides a day. EWG has an excellent shopper's guide that you can download from its website.

"There are so many different food labels—organic, natural—what do they all mean?"
Here is a rundown of food-label terms every parent should know:

- *100% Organic:* These fruits and vegetables are farmed using only botanical or nonsynthetic pest-control methods.

- *Organic:* At least 95 percent of the ingredients in these products are organically produced. However if 80 percent of a chicken's feed is organically grown, it gets the "organic" sticker. Seafood is an exception because the USDA has not yet established a standard. Antibiotics and other drugs may be used for animals in organic farming in certain circumstances.

- *Made with Organic Ingredients:* This label means that 70 percent of the ingredients in the food product are organic.

- *Free Range or Free Roaming:* These labels, which are interchangeable, can be found on poultry and eggs. They let you know that the door of each chicken's cage was opened for an unspecified period of time each day.

- *Natural or All Natural:* There are no government restrictions nor is there a specific definition for this label. Producers and manufacturers of food may simply decide whether or not they want to use it. Although it implies that the food product has no artificial ingredients and is minimally processed, there are no set standards.

Study: Household pesticides linked to childhood leukemia

From 1995 to 1999 Dr. Xiaomei Ma and a team of researchers studied the correlation between pesticide exposure and childhood leukemia in northern California. By interviewing the parents of eighty-one newly diagnosed kids, who ranged in age from birth to fourteen years, and comparing their family pesticide exposure with that of eighty-one healthy children, Dr. Ma was able to analyze the effects of pesticide usage during three key developmental periods: three months before pregnancy, during pregnancy, and each year after birth for three consecutive years. Dr. Ma found that children who have leukemia are twice as likely to have been exposed to pesticides applied by professional pest control services. The study also showed that risk of leukemia is significantly elevated if the application occured when a child was in utero.

"What is the safest way to rid my house of bugs and rodents?"

In 2000 the EPA banned organophosphates, one of the pesticides most dangerous to human health, as a household pesticide. Since the ban, levels of organophosphates in people's bodies have dropped dramatically. However, pesticides for use inside our homes contain chemicals that have not been thoroughly researched for effects in children. Therefore, it is advisable to use nonchemical methods, called integrated pest management (IPM), for deterring insects and rodents.

Integrated Pest Management (IPM)

Given what little we know about how chemicals designed to kill pests affect our children's health, it is important to use nonchemical alternatives whenever possible. Integrated pest management (IPM), a multipronged approach, is the best means for keeping insects and rodents out of your home and garden with the least use of hazardous chemicals. Here are a few easy things to do:

- Make your house unappealing to insects and rodents:
  - Sponge down spills—food and sugary drinks.
  - Eliminate clutter around your house so that pests don't build nests.
  - Keep food in sealed containers.
  - Don't leave food out overnight.
  - Be aware that pet food can also attract bugs and rodents.

- Keep your home well maintained.
  - Repair leaky faucets or pipes to keep pests from making their home near a good water source.
  - Block all holes into your home, especially in the kitchen, by caulking cracks and crevices in floors and foundations.
  - Look under your sinks where pipes pass through walls and floors and make sure there are no gaps. I use steel wool and insulation to close off possible passageways for mice, rats, and other rodents.

- Set traps.
  - Use traps at night when mice and other rodents are most active. I use a small amount of peanut butter

157

on the trap for bait. Set traps about five to ten feet apart near rodent activity.

- Use the least toxic alternatives.
  - If all else fails, use nontoxic pesticides. Go to Beyond Pesticides at www.beyondpesticides.org for additional information about pesticides free of hazardous chemicals.

If none of these suggestions work, consult a pest control specialist who uses IPM methods for controlling pests.

"Is it safe to use pesticides and herbicides on my lawn and garden?"

Some lawn and garden chemicals have been linked to brain and nervous system injuries in rodent studies. When you put these hazardous chemicals on your lawn they can end up on your children and on your carpets and sofas when they are tracked in on the bottoms of shoes. Chemicals used on your lawn and in your garden can also end up in your groundwater.

WHAT YOU CAN DO TO
PROTECT CHILDREN FROM
PESTICIDES

- Feed your children organic food, especially for those foods that are otherwise grown with high levels of pesticides (page 153 lists low- and high-pesticide-residue produce). For a complete list of foods' pesticide levels, go to www.foodnews.org. For a list of local food growers all over the country, go to www .ams.usda/farmersmarket, www.sare.org, and www .localharvest.org.

- Buy locally or order organic meat and poultry. There are several websites to order from www.eatwild .com, www.mynaturalbeef.com, www.freshdirect .com, www.eatwellguide.org, and www.theorganic pages.com.

- If you have a pest problem in your home, follow the

integrated pest management practices outlined on page 156.

- Avoid using chemicals on your lawn. Instead, leave grass clippings on your grass after you mow. It acts like compost, breaking down and releasing nitrogen into the soil.

- Never cut your lawn shorter than three inches. This allows root systems to become strong and also pushes out weeds.

FOR FURTHER INFORMATION

Hospitals for a Healthy Environment: www.h2e-online.org

Beyond Pesticides: www.beyondpesticides.org

Washington's Toxic Coalition: www.watoxics.org

Northwest Alternative to Pesticides: www.pesticide.org

Healthy Child Healthy World: http://healthychild.org

Information on food labels: www.eco-labels.org

8

A SAFER FUTURE

WHY THERE'S REASON TO HOPE— AND TO STAY INFORMED

The last twelve months have been, in many ways, like leaping into the swimming hole down the road a ways from my home in the Adirondacks. Every time I jump from the cliff at Little Falls, I take deep breaths before hurling myself off the rocks, making sure to jump out far enough to clear the ledge below. Even so, it is still a shock when my feet smack against the cold water. Sinking down into the mud, I look up and it is nearly impossible to make out the shadowy shapes above the waterline. Similarly, when I started this book I felt as though I was at the bottom of a deep hole, filled with a dense clump of mystifying factoids and perceptions knit together by anxiety. However, as I really began to look at the science and talk to the experts, clear patterns began to emerge and I eventually broke through the overwhelming apprehension I felt about environmental toxins and our children's health.

Today I am relieved. There are at least three good reasons for hope. First, I am encouraged and inspired by the numbers of scientists I have spoken to in the past year who have committed their lives to understanding how toxins injure children. Their research has the power to create change at both the legislative and the corporate level. For instance, laws that banned lead and PCBs only came about because the scientific community reached a consensus that these substances were harmful. And this process continues. After Thomas Zoeller of the University of Massachusetts published a study in *Endocrinology* (February 2005) that showed how bisphenol A hinders the ability of thyroid hormone to regulate proper in utero brain development in lab animals, Assemblywoman Wilma Chan (D-Oakland) introduced the Stop Toxic Toys bill in California. The bill called for prohibiting the sale of any toy that contains bisphenol A or phthalates for use on a child under three. Although the bill did not pass at the state level, the San Francisco Board of Supervisors unanimously adopted the ordinance in June 2006. Fearful that the new legislation would hurt sales, retailers and industry lobby groups (Citikids Baby News, American Chemistry Council, California Retailers Association, California Grocers Association, and Juvenile Products Manufacturer Association) filed a lawsuit against San Francisco's ban. Even so, the stab at changing laws to protect small children from these hormone-disrupting chemical compounds made people aware of the threat. In December 2006 *TIME* Magazine ran an article about the evils of chemicals in plastic, and the legislation in California will not only be reintroduced in that state, but scientists will be asked to testify in other states as well, where similar legislation has been proposed.

Second, more and more companies are taking the initiative to eliminate harmful chemicals from their products before they are required to do so by law. Some of the forerunners include Ikea, Shaw Carpets, Volvo, Dell, Patagonia, and IBM. Momentum continues to build as other companies join the movement toward environmental responsibility.

Third, I take comfort that current trends in the food industry show that we as consumers have a substantial say about the chemicals that are used to grow and manufacture our food. Concerned that pesticides are contributing to life-threatening illnesses, more and more Americans are buying organic and natural food. The National Grocers Association (NGA) reports that natural foods are now the fastest-growing product area in supermarkets, with conventional food companies like Gerber and Frito-Lay joining the trend. The NGA estimates that by the year 2010, sales of organic food will reach $22 billion. Their findings have also shown that the demographic profile of organic shoppers is widening. People in all geographic areas, age groups, income levels, and education levels are purchasing organic products. Demand is so great across the board that even Wal-Mart recently added an organic section.

On the other hand, given the pace at which new chemicals enter the marketplace (in the United States about 1,500 new chemicals are unleashed into the environment every year), this book will probably have to be written all over again before today's schoolchildren graduate. Even though many poisonous compounds like PCBs and organophosphate pesticides for home use have been yanked off the market, I am dismayed that there always seems to be another substance out there poised to take their place as a threat to the health and proper development of our children.

For instance, a family of chemical compounds called perfluorochemicals (PFCs)—used in fast-food wrappers; in microwave popcorn bags; as a stain repellent on clothes, carpets, and upholstery; and in nonstick frying pans—has recently emerged as a toxic substance. Scientists believe PFCs are constantly leaching from the surfaces they are applied to and filtering into indoor air. How much is safe? Just as with so many of the toxins discussed in this book, no one knows for sure. However, scientists have established that PFCs remain in the environment and in our bodies for a long time, and people in the United States have far more PFCs in their blood than any other population on the globe.

In another twenty years my children will be grown and I will be thinking about the health and well-being of my grandchildren. Last summer at an Adirondack farmers' market, I caught a glimpse of a brand-new baby. His tiny face, partially hidden inside a woven Guatemalan sling, was squished and pink. Although my youngest is only three, I already miss the smell of a young baby's warm scalp, the certain grip of a tiny fist closing around my finger. I was lost for a moment, thinking about babies, when I felt a tug at my arm. I turned my head and focused my gaze on Matilda, our eleven-year-old, standing next to me with a mesh bag crammed with spinach slung over her shoulder. She caught my eye, then pointed at a toddler in a stroller. In a tie-dyed onesie, the child, who was a sticky mess, was devouring a peach. A rivulet of pulp ran down his chin and he forcefully stuck out his tongue, swiveling it around his lips.

As we drove the hilly mountain roads heading home, Matilda chattered, cataloging the host of additions she intended to make to her wardrobe before starting junior high in the fall, and my mind began to wander. I smiled as I thought about her reaction to the little boy. Then, even though I knew it was a long way off, I found myself thinking that there is a possibility that the world will be different by the time Matilda has her own babies. That maybe, by then, things will have gotten a lot better for children. That perhaps there is a chance, if intrepid mothers and fathers, public health scientists, lawmakers, and corporate leaders pay close enough attention to the kind of chemicals we are dispersing in the environment, that our children's children will have the opportunity to play with toys manufactured without hormone-mimicking chemicals and to nibble tuna sandwiches free of trace amounts of neurotoxins. With vigilant parents, consumers, and researchers all working toward this goal, there is reason to hope.

TEN KEY STEPS

Follow these ten key steps and you can pretty much be assured that you are effectively protecting your child from environmental toxins.

1. If you live in a house built before 1978, have your pediatrician test your young children for lead, and ask for the results. If your child's lead count is even as high as two milligrams per deciliter, scour your home for possible sources of lead. Likely candidates include chipping paint inside and outside your home, as well as on any window frames, painted dishes, and painted antiques. Even my mother-in-law's baby chair turned out to be a lead culprit.

2. If you're intending to become pregnant, are already pregnant, or are preparing fish for young children, consult the list of low-mercury fish on page 60 and stick to the least-contaminated varieties. Keep in mind that some public health scientists suggest that women of childbearing age and young children should avoid fish altogether.

3. Take a hard look at all plastics in your house. Throw out the soft plastics (for instance, shower curtains and wash-off baby books) that are made with phthalates and get rid of shiny, hard baby bottles and water bottles that might contain bisphenol A.

4. Eliminate persistent industrial contaminants like PCBs from your child's diet by serving low-fat meats and avoiding farmed fish, particularly "Atlantic" salmon.

5. Avoid flame retardants by purchasing furniture, children's pajamas, electronics, cars, and carpets from companies that state clearly that they manufacture products free of toxic chemicals. Consult *The Green Guide* at www.greenguide.com for a list of environmentally responsible manufacturers.

6. If school buses, minivans, cars, and trucks are idling outside your child's school, enlist the help of other parents, the school administration, the school board, and the bus company to create a no-idling policy.

7. Keep your children inside during peak hours if the radio says ozone is high.

8. Reduce pesticides in your food by buying the following produce in the organic section: apples, bell peppers, celery, cherries, imported grapes, nectarines, peaches, pears, potatoes, red raspberries, spinach, and strawberries.

9. Be careful in dealing with pests. Use integrated pest management as outlined on page 156.

10. Visit www.toxicsandbox.com to stay current on environmental threats to our children.

QUESTIONS FOR YOUR PEDIATRICIAN AND DENTIST

1. If you live in a home built before 1978, ask your pediatrician to check your child's lead count by the time he or she is one. Take the actions described in this book if your child's level measures above two milligrams per deciliter.

2. Ask for vaccines that are free of the preservative thimerosal (which contains ethyl mercury). This has been eliminated from most vaccines. Your doctor should be able to find nonthimerosol vaccines.

3. If your child gets a cavity, ask your dentist for an alternative to the silver-colored amalgam filling that contains mercury.

ABOUT THE BLOG

Visit the Toxic Sandbox blog at www.toxicsandbox.com.

Keep abreast of up-to-the-minute research, meet other parents online, download lists of safe products, participate in conversations with the experts, and find links to other sites committed to the healthy development of our children.

NOTES

NOTES

180

SOURCES

Below is a list of the primary sources that informed my research and conclusions about children's health and environmental toxins. I encourage you to consult these and other primary sources as you continue to stay informed.

CHAPTER 1: UNCOVERING THE TRUTH
To find good information about what chemicals have been detected in our bodies go to www.ewg.com.

Gore, Al. *An Inconvenient Truth: The Planetary Emergency of Global Warming and What We Can Do about It.* New York: Rodale, 2006.

Landrigan, Phillip, et al. "Environmental Pollutants and Disease in American Children: Estimates of Morbidity, Mortality, and Costs for Lead Poisoning, Asthma, Cancer, and Developmental Disabilities." *Environmental Health Perspectives* 100, no. 7 (July 2002).

Lanphear, Bruce P. "Origins and Evolution of Children's Environmental Health." *Environmental Health Perspectives* 113 (2005).

Steingraber, Sandra. *Having Faith: An Ecologist's Journey to Motherhood*. New York: Berkley, 2003. This book, which informed so much of my thinking about children and pollutants, is a must-read for anyone wanting more information on how toxins can injure our children. It served as an indispensable resource for my chapters on lead, mercury, and PCBs.

Trasande, Leo, et al. "The Environment in Pediatric Practice: A Study of New York Pediatricians' Attitudes, Beliefs, and Practices towards Children's Environmental Health." *Journal of Urban Health: Bulletin of the New York Academy of Medicine* (2006).

Trasande, Leo, et al. "Pediatrician Attitudes, Clinical Activities, and Knowledge of Environmental Health in Wisconsin." *Wisconsin Medical Journal* 105, no. 2 (2006).

CHAPTER 2: LEAD

I found the Healthy Child Healthy World website (http://healthychild.org) to be informative about lead and children's health.

Braun, Joe, et al. "Exposures to Environmental Toxicants and Attention Deficit Hyperactivity Disorder in US Children." *Environmental Health Perspectives* 114, no. 12 (December 2006).

Canfield, Richard, et al. "Intellectual Impairment in Children with Blood Lead Concentrations below 10 µg per Deciliter." *New England Journal of Medicine* 348, no. 16 (April 17, 2003).

Gulson, Brian, et al. "Blood Lead Changes during Pregnancy and Postpartum with Calcium Supplementation." *Environmental Health Perspectives*. 12, no. 15 (November 2004).

Gulson, Brian, et al. "Mobilization of Lead from Human Bone Tissue during Pregnancy and Lactation—a Summary of Long-Term Research." *Science Total Environment* 303 (2003): 79–104.

Gulson, Brian, et al. "Pregnancy Increases Mobilization of Lead from Maternal Skeleton." *The Journal of Laboratory and Clinical Medicine* 130 (July 1997): 51–62.

Landrigan, Philip J., et al. *Raising Healthy Children in a Toxic World.* New York: Rodale, 2001.

Lanphear, Bruce, et al. "Low-Level Environmental Lead Exposure and Children's Intellectual Function: An International Pooled Analysis." *Environmental Health Perspectives* 113, no. 7 (July 2005).

Lanphear, Bruce, et al. "Cognitive Deficits Associated with Blood Concentrations Less than 10 µg/dl in US Children and Adolescents." *Public Health Reports* 115 (November/December 2000).

Markowitz, Gerald, and David Rosner. *Deceit and Denial: The Deadly Politics of Industrial Pollution.* Berkeley and Los Angeles: University of California Press, 2002. A great read on the history of lead.

Needleman, Herbert. "Childhood Lead Poisoning: The Promise and Abandonment of Primary Prevention." *American Journal of Public Health* 88, no. 12 (December 1998): 1871–76.

Rapp, Doris. *Is This Your Child's World?* New York: Bantam, 1996.

Schettler, Ted, et al. *Generations at Risk: Reproductive Health and the Environment.* Cambridge, MA: MIT

Press, 1999. This book informed much of my thinking not only on lead but also on mercury, phthalates, bisphenol A, and PCBs.

Schneider, Dona, and Freeman, Natalie. *Children's Environmental Health: Reducing Risk in a Dangerous World*. Washington, DC: American Public Health Association, 2000.

Schneider, J. S., et al. "Enriched Environment during Development Is Protection against Lead-Induced Neurotoxicity." *Brain Research* (March 2001).

Steingraber, Sandra. *Having Faith: An Ecologist's Journey to Motherhood*. (New York: Berkley, 2003).

CHAPTER 3: MERCURY

Good websites for lists of mercury content in fish are the Environmental Working Group (www.ewg.com) and *The Green Guide* (www.thegreenguide.com).

Bellinger, David, et al. "Neuropsychological and Renal Effects of Dental Amalgam in Children: A Randomized Trial." *Journal of the American Medical Association* 295, no. 15 (April 2006).

Davidson, Philip W., et al. "Mercury Exposure and Child Development Outcomes." *Pediatrics* 113, no. 4 (April 4, 2004): 1023–29.

Grandjean, P., et al. "Cognitive Deficit in 7-Year-Old Children with Prenatal Exposure to Methylmercury." *Journal of Pediatrics* (February 2004).

Myers, G. J., et al. "Prenatal Methylmercury Exposure from Ocean Fish Consumption in the Seychelles Child Development Study." *The Lancet* 361 (May 17, 2003).

Saito, Hisashi. "Congenital Miyama Disease: A Description of Two Cases in Niigata." *Seychelles Medical and Dental Journal* 7, no. 1 (November 2004).

Schettler, Ted, et al. *In Harm's Way: Toxic Threats to Child Development.* A report by Greater Boston Physicians for Social Responsibility (May 2000).

Schettler, Ted, et al. *Generations at Risk: Reproductive Health and the Environment.* Cambridge, MA: MIT Press, 1999.

Steingraber, Sandra. *Having Faith: An Ecologist's Journey to Motherhood.* New York: Berkley, 2003.

Trasande, Leo, et al. "Public Health and Economic Consequences of Methylmercury Toxicity to the Developing Brain." *Environmental Health Perspectives* 113, no. 5 (May 2005).

CHAPTER 4: PLASTICS

Colburn, Theo, Dianne Dumanoski, and John Peterson Myers. *Our Stolen Future: Are We Threatening Our Fertility, Intelligence, and Survival?* New York: Plume, 1997. This is the book that was particularly informative in my thinking about endocrine disruptors and children's health.

Howdeshell, Kembra, and Frederick Vom Saal. "Developmental Exposure to Bisphenol A: Interaction with Endogenous Estradiol during Pregnancy in Mice." *American Zoology* 40 (2000): 429–37.

Rapp, Doris. *Is This Your Child's World?* New York: Bantam, 1996.

Schettler, Ted, et al. *Generations at Risk: Reproductive Health and the Environment.* Cambridge, MA: MIT Press, 1999.

Swaw, Shanna. "Decrease in Anogenital Distance among Male Infants with Prenatal Phthalate Exposure." *Environmental Health Perspectives* 113, no. 8 (August 2005).

Zoeller, R. T., et al. "Bisphenol-A, an Environmental Contaminant That Acts as a Thyroid Hormone Receptor Antagonist In Vitro, Increases Serum Thyroxin, and

Alters RC3/Neurogranin Expression in the Developing Rat Brain." *Endocrinology* (February 2005).

CHAPTER 5: PCBs AND FLAME RETARDANTS

The Environmental Working Group (www.ewg.com) did a good analysis of dust in homes across the country that is worth consulting if you want additional information about how these toxins are finding their way into our bodies.

A fantastic website for general information about PCBs and PBDEs is Healthy Child Healthy World (http://healthychild.org).

Alm, Henrik, et al. "Protemic Evaluation of Neonatal Exposure to 2, 2, 4, 4, 5–Pentabromodiphenyl Ether." *Environmental Health Perspectives* 114, no. 2 (February 2006).

Eriksson, Per, et al. "Brominated Flame Retardants: A Novel Class of Developmental Neurotoxicants in Our Environment?" *Environmental Health Perspectives* 109, no. 9 (September 2001).

Fischer, Douglas, et al. "Children Show Highest Levels of Polybrominated Diphenyl Ethers in a California Family of Four: A Case Study." *Environmental Health Perspectives* 114, no. 10 (October 2006).

Hites, Ronald, et al. "Global Assessment of Organic Contaminants in Farmed Salmon." *Science* 303, no. 5655 (January 2004): 226–29.

Jacobson, Joseph, et al. "Effects of In Utero Exposure to Polychlorinated Biphenyls and Related Contaminants on Cognitive Functioning in Young Children." *Journal of Pediatrics* (January 1990).

Lai, T. J., et al. "A Cohort Study of Behavioral Problems and Intelligence in Children with High Prenatal Polychlorinated Biphenyl Exposure." *Archives of General Psychiatry* (November 2002).

Mazdai, Anita, et al. "Polybrominated Diphenyl Ethers in Maternal and Fetal Blood Samples." *Environmental Health Perspectives* 111, no. 9 (July 2003).

Schecter, Arnold, et al. "Polybrominated Diphenyl Ethers (PBDEs) in US Mothers' Milk." *Environmental Health Perspectives* 111, no. 14 (November 2003).

Schettler, Ted, et al. *Generations at Risk: Reproductive Health and the Environment.* Cambridge, MA: MIT Press, 1999.

Steingraber, Sandra. *Having Faith: An Ecologist's Journey to Motherhood.* New York: Berkley, 2003.

Stewart, Paul, et al. "Prenatal PCB Exposure and Neonatal Behavioral Assessment Scale (NBAS) Performance." *Neurotoxicology and Teratology* 22 (2000): 21–29.

Washam, Cynthia. "Concentrating on PBDEs: Chemical Levels Rise in Women." *Environmental Health Perspectives* (July 2003).

Williams, Florence. "Toxic Breast Milk?" *The New York Times Magazine* (January 9, 2005).

CHAPTER 6: AIR POLLUTION
Some of the best research about air pollution and children's health is coming from Columbia University's Center for Children's Environmental Health at the Mailman School of Public Health. You can access these studies at its website (www.mailman.hs.columbia.edu).

American Lung Association. "State of the Air: 2006." www.lungusa.org/reports/stateoftheair2006.

———. "State of the Air: 2004." www.lungaction.org/reports/stateoftheair2004.

Gauderman, James, et al. "Air Pollution Retards Teen Lung Growth." *New England Journal of Medicine* (September 2004).

Gore, Al. *An Inconvenient Truth: The Planetary Emergency of Global Warming and What We Can Do about It.* New York: Rodale, 2006.

Kim, Janice. "Traffic-Related Air Pollution near Busy Roads." *American Journal of Respiratory and Critical Care Medicine* 170 (2004): 520–26.

Perera, Frederica, et al. "Effects of Prenatal Exposure to Airborne Polycyclic Aromatic Hydrocarbons on Neurodevelopment in the First Three Years of Life among Inner-City Children." *Environmental Health Perspectives* 114, no. 8 (August 2006).

Samet, Jonathan, et al. "The National Morbidity, Mortality, and Air Pollution Study." *Health Effects Institute,* no. 94, Parts I and II (June 2000).

Wang, X., et al. "Association between Air Pollution and Low Birth Weight: A Community-Based Study." *Environmental Health Perspectives* 105, no. 5 (May 1997).

Wargo, John. "Children's Exposure to Diesel Exhaust on School Buses." *Environmental and Human Health* (February 2002).

Woodruff, Tracey, et al. "The Relationship between Selected Causes of Postneonatal Infant Mortality and Particulate Air Pollution in the United States." *Environmental Health Perspectives* 105, no. 6 (June 1997).

Zmirou, D., et al. "Traffic-Related Air Pollution and Incidence of Childhood Asthma: Results of the Vesta Case-Control Study." *Journal of Epidemiology and Community Health* 58 (2004): 18–23.

CHAPTER 7: PESTICIDES
The Environmental Working Group provides useful information at its website (www.ewg.com).

Daniels, Julie, et al. "Pesticides and Childhood Cancers." *Environmental Health Perspectives* 105, no. 10 (October 1997).

Eskenazi, Brenda. "Exposures of Children to Organophosphate Pesticides and Their Potential Adverse Health Effects." *Environmental Health Perspectives* 107, no. S3 (June 1999).

Guillette, Elizabeth, et al. "An Anthropological Approach to the Evaluation of Preschool Children Exposed to Pesticides in Mexico." *Environmental Health Perspectives* 106, no. 6 (June 1998).

Landrigan, Philip J., et al. *Raising Healthy Children in a Toxic World*. New York: Rodale, 2001.

Lu, Chensheng, et al. "Organic Diets Significantly Lower Children's Dietary Exposure to Organophosphorous Pesticides." *Environmental Health Perspectives* 114, no. 2 (February 2006).

Ma, Xiaomei, et al. "Critical Windows of Exposure to Household Pesticides and Risk of Childhood Leukemia." *Environmental Health Perspectives* 110, no. 9 (September 2002).

Rapp, Doris. *Is This Your Child's World?* New York: Bantam, 1996.

Schettler, Ted, et al. *Generations at Risk: Reproductive Health and the Environment.* Cambridge, MA: MIT Press, 1999.

Zahm, S. H. "Childhood Leukemia and Pesticides." *Epidemiology* 10, no. 5 (September 1999).

Zahm, S. H., and Mary Ward. "Pesticides and Childhood Cancer." *Environmental Health Perspectives* 106, no. S3 (June 1998).

CHAPTER 8: A SAFER FUTURE

McDonough, William, and Michael Braungart. *Cradle to Cradle: Remaking the Way We Make Things.* New York: North Point Press, 2002.

ACKNOWLEDGMENTS

This book could not have been written without the constant support given to me by my three older children, Polly, Tilly, and Daniel—babysitting Super George and doing double duty with chores. Then there were the dozens of mothers, who spoke to me multiple times about their concerns regarding environmental toxins: Cena Shaw, Meg Niles, Kristen Fiegl, Chelly Hegan, Cynthia Poppino, Jennifer McDonald, and Darla Breckenridge. My dear friend Leslie Fram, who guided me every step of the way. Jennifer Brown, who spent hundreds of hours on the phone with me, enlightening me about the human body. Judith Enck, who gave me the courage to take on this project and introduced me to public health scientists. Patty Goodwin, who thought up the title and edited my book proposal. Gareth Esersky, who advised and supported me every step of the way—since the afternoon we met on the beach in Dominican Republic her direction has been crucial to the actualization of this book. Marion Roach, my writing teacher, who encouraged me from the beginning. Claudine Paris, who lovingly edited the entire manuscript. David Carpenter, Ted Schettler, Elizabeth Guillette, Ricky Perera, Leo Trasande, Fred Vom Saal, and all the other public health scientists who gave me their valuable time. My team of overqualified babysitters, who were so smart and creative in their care of my children that I never had to worry. Each of them—Nayeli Rodriguez, Al Sand, and Krissy Caggiano—quickly climbed into our hearts and became part of our family. Finally, David, whose constant support of this book, which included an astonishing number of Sundays devoted to taking the children on various excursions so I could be alone at my computer, buoyed me along throughout this year of research and writing.

INDEX

Page numbers in *italics* indicate illustrations.

ABOUT THE AUTHOR

Libby McDonald is a writer and documentary filmmaker who lives with her husband and four children in the Adirondack Mountains. Her film credits include two investigative documentary films, *New School Order* (Sundance Film Festival Official Entry) and *Terror Town*. An advocate for children, Ms. McDonald has started two schools, Learning Community, an alternative public school for underprivileged children in Jersey City, New Jersey, and Lakeside Preschool, a Waldorf-inspired preschool in upstate New York.